T0184028

Communications
in Computer and Information Science 1003

Commenced Publication in 2007
Founding and Former Series Editors:
Phoebe Chen, Alfredo Cuzzocrea, Xiaoyong Du, Orhun Kara, Ting Liu,
Krishna M. Sivalingam, Dominik Ślęzak, Takashi Washio, and Xiaokang Yang

Editorial Board Members

Simone Diniz Junqueira Barbosa
  Pontifical Catholic University of Rio de Janeiro (PUC-Rio),
  Rio de Janeiro, Brazil
Joaquim Filipe
  Polytechnic Institute of Setúbal, Setúbal, Portugal
Ashish Ghosh
  Indian Statistical Institute, Kolkata, India
Igor Kotenko
  St. Petersburg Institute for Informatics and Automation of the Russian
  Academy of Sciences, St. Petersburg, Russia
Junsong Yuan
  University at Buffalo, The State University of New York, Buffalo, NY, USA
Lizhu Zhou
  Tsinghua University, Beijing, China

More information about this series at http://www.springer.com/series/7899

Yannis Manolopoulos · Sergey Stupnikov (Eds.)

Data Analytics and Management in Data Intensive Domains

20th International Conference, DAMDID/RCDL 2018
Moscow, Russia, October 9–12, 2018
Revised Selected Papers

Springer

Editors
Yannis Manolopoulos ⓘ
Open University of Cyprus
Latsia, Cyprus

Sergey Stupnikov ⓘ
Federal Research Center "Computer Science
and Control" of RAS
Moscow, Russia

ISSN 1865-0929 ISSN 1865-0937 (electronic)
Communications in Computer and Information Science
ISBN 978-3-030-23583-3 ISBN 978-3-030-23584-0 (eBook)
https://doi.org/10.1007/978-3-030-23584-0

© Springer Nature Switzerland AG 2019
This work is subject to copyright. All rights are reserved by the Publisher, whether the whole or part of the material is concerned, specifically the rights of translation, reprinting, reuse of illustrations, recitation, broadcasting, reproduction on microfilms or in any other physical way, and transmission or information storage and retrieval, electronic adaptation, computer software, or by similar or dissimilar methodology now known or hereafter developed.
The use of general descriptive names, registered names, trademarks, service marks, etc. in this publication does not imply, even in the absence of a specific statement, that such names are exempt from the relevant protective laws and regulations and therefore free for general use.
The publisher, the authors and the editors are safe to assume that the advice and information in this book are believed to be true and accurate at the date of publication. Neither the publisher nor the authors or the editors give a warranty, expressed or implied, with respect to the material contained herein or for any errors or omissions that may have been made. The publisher remains neutral with regard to jurisdictional claims in published maps and institutional affiliations.

This Springer imprint is published by the registered company Springer Nature Switzerland AG
The registered company address is: Gewerbestrasse 11, 6330 Cham, Switzerland

Leonid Kalinichenko

10.06.1937–17.07.2018

In July 2018 we lost a leading visionary scientist in the field of the database theory. Leonid Kalinichenko is known in the scientific community for his pioneering works, the scientific school he established, and the formation of two international scientific conferences run for two decades each.

Leonid Kalinichenko's entire life was devoted to computer science. In 1959 he graduated from Kiev Polytechnic Institute and started his work at the Institute of Cybernetics the Ukraine Academy of Sciences. In 1968 he received his Ph.D. degree from the Institute of Cybernetics the main results of his thesis were devoted to discrete events systems simulation (languages, tools, applications), and in 1969 he moved to the Institute for Electronic Control Machines, Moscow. Later, in 1985, he received his Doctor of Sciences degree from the Lomonosov Moscow State University. The thesis was devoted to methods and tools of heterogeneous databases integration. In the same year he became the head of a department at the Institute of Informatics Problems, Academy of Sciences of the USSR, which is now the Federal Research Center "Computer Science and Control" of the Russian Academy of Sciences (FRC CSC RAS).

In 1978 Leonid joined the faculty of the Lomonosov Moscow State University and in 1990 he became Professor of the Department of Computational Mathematics and Cybernetics. He taught courses on object-oriented databases and distributed object technologies. He is an author of ten books, and more than 200 research papers in journals and conference proceedings. His research interests included interoperable heterogeneous information resource integration and mediation, semantic interoperability, compositional development of information systems, middleware architectures, and digital libraries.

Several significant program systems were developed under his supervision: discrete event system simulation SLANG (1969), embedding of simulation systems into various programming systems SKIF (1977), heterogeneous database integration (1984), compositional development of interoperable information systems (2001), subject mediation middleware for scientific problem solving over distributed information resources

(2010). Investigations on methods and tools for heterogeneous information resources integration applying subject mediation methodology supported by formal specification and verification were embraced by the framework project called SYNTHESIS. In 1986 Leonid Kalinichenko was awarded the USSR State Prize in the Field of Science and Technology for his works on system simulation. In 2010 he became Honored Scientist of the Russian Federation.

For several decades Leonid Kalinichenko took an active part in facilitation of research in the field of databases. From 1974 he was the Deputy Chairman of the Working Group on Software for Data Banks under the USSR State Committee on Science and Technology. He initiated the creation of the Moscow ACM SIGMOD Chapter in 1992 and became the permanent chair of the chapter. The monthly scientific seminar of the chapter has operated since the chapter creation until the present time. The 200 seminar meetings significantly influenced the Russian community on databases and information systems.

He successfully formed several international conferences including the European Conference on Advances in Databases and Information Systems (ADBIS) in 1993 and Russian Conference on Digital Libraries (RCDL) in 1999. Leonid Kalinichenko acted as the permanent chair of the Steering Committees of the conferences as well as chair (co-chair) of the Program Committees of many conferences.

In the last few years his activities and works were devoted to problem solving in data intensive domains. During 2013–2016 he initiated several research projects aimed at conceptual modeling and data integration within distributed computational infrastructures.

In 2015 he initiated the transformation of the RCDL conference into the International Conference on Data Analytics and Management in Data Intensive Domains (DAMDID)—a multidisciplinary forum of researchers and practitioners from various domains of science and research promoting the cooperation and exchange of ideas in the area of data analysis and management in data-intensive domains. The conference became a place for discussions on data access, analysis and management problems in astronomy, neurology, genomics, material science, biology.

In 2015 he also organized a master program entitled "Big data: infrastructures and methods for problem solving" at the Department of Computational Mathematics and Cybernetics, Lomonosov Moscow State University to attract students in the field of multidisciplinary data analysis and management.

Leonid was a great scientist with a surprisingly deep knowledge of the state of the art in his field of science, understanding the urgent directions of the development of science. But he was also a real scientific driver of the research team he led at the Institute of Informatics Problems, a driver of conferences he established and scientific groups he contacted.

Leonid's passing away is a huge loss for the scientific community, his family, his colleagues and his friends.

Igor Sokolov
Yannis Manolopoulos
Vladimir Sukhomin
Victor Zakharov
Nikolay Kolchanov
Arkady Avramenko
Pavel Braslavsky
Vasily Bunakov
Alexander Elizarov
Alexander Fazliev
Alexei Klimentov
Mikhail Kogalovsky
Vladimir Korenkov
Mikhail Kuzminski
Sergey Kuznetsov
Vladimir Litvine
Archil Maysuradze
Oleg Malkov
Alexander Marchuk
Igor Nekrestjanov
Boris Novikov
Nikolay Podkolodny
Aleksey Pozanenko
Vladimir Serebryakov
Yury Smetanin
Vladimir Smirnov
Sergey Stupnikov
Konstantin Vorontsov
Viacheslav Wolfengagen

Preface

This CCIS volume published by Springer contains the proceedings of the XX International Conference Data Analytics and Management in Data-Intensive Domains (DAMDID/RCDL 2018) that took place during October 9–12 in the Lomonosov Moscow State University at the Department of Computational Mathematics and Cybernetics. The conference was dedicated to the memory of its founder, Leonid Kalinichenko, who passed away on July 17, 2018.

The DAMDID is planned as a multidisciplinary forum of researchers and practitioners from various domains of science and research, promoting cooperation and exchange of ideas in the area of data analysis and management in domains driven by data-intensive research. Approaches to data analysis and management being developed in specific data-intensive domains (DID) of X-informatics (such as X = astro, bio, chemo, geo, med, neuro, physics, chemistry, material science etc.), social sciences, as well as in various branches of informatics, industry, new technologies, finance and business are expected to contribute to the conference content.

Traditionally DAMDID/RCDL proceedings are published locally before the conference as a collection of full texts of all contributions accepted by the Program Committee: regular and short papers, abstracts of posters and demos. Soon after the conference, the texts of regular papers presented at the conference are submitted for online publishing in a volume of the European repository of the CEUR Workshop Proceedings, as well as for indexing the volume content in DBLP and Scopus. Since 2016 a DAMDID/RCDL volume of post-conference proceedings with up to one third of the submitted papers that have been previously published in CEUR Workshop Proceedings have been published by Springer in their Communications in Computer and Information Science (CCIS) series. Each paper selected for the CCIS post-conference volume should be modified as follows: the title of each paper should be a new one; the paper should be significantly extended (with at least 30 per cent of new content); the paper should refer to its original version in CEUR Workshop Proceedings.

The program of DAMDID/RCDL 2018 alongside the traditional data management topics reflects a rapid move into the direction of data science and data-intensive analytics.

The Workshop on FAIR Data and European Open Science Cloud (EOSC) constituting the first day of the conference on October 9 included four sessions. The first session devoted to FAIR (Findable, Accessible, Interoperable, Reusable) data principles in the realm of open science included invited talks by Michel Schouppe (European Commission, Directorate-General for Research and Innovation) and Simon Hodson (Executive Director of CODATA). Michel Schouppe considered the relevance of EOSC and FAIR and phases of implementing the EOSC, while Simon Hodson overviewed FAIR Data, FAIR Services and the FAIR data action plan. The second section devoted to FAIR platforms and interoperability included invited talks by Ari Asmi

(Integrated Carbon Observation System ERIC), Sergey Stupnikov (FRCCSC RAS) and Nikolay Skvortsov (FRCCSC RAS). Ari Asmi discussed building FAIR environmental services platforms in Europe, Sergey Stupnikov presented FAIR data based on extensible unifying data model development, and Nikolay Skvortsov considered data interoperability and reuse among heterogeneous scientific communities. The third session devoted to FAIR implementation issues and activities included invited talks by Peter Wittenburg (Max Planck Computing and Data Facility), Erik Schultes (Leiden University Medical Centre, GO FAIR International Support and Coordination Office) and Damien Lecarpentier (CSC-IT Center for Science). Peter Wittenburg discussed digital objects as a concept to help implement FAIR and EOSC, Erik Schultes considered accelerating convergence to an Internet of FAIR data and services, and Damien Lecarpentier overviewed FAIR activities within the various EOSC-funded initiatives. The fourth session was a panel on data access challenges for data-intensive research in Russia and EOSC where experts from different regions and disciplines had the chance to discuss the current state and future of the FAIR action plan and the EOSC initiative, present their views and contributions, and discuss active participation.

The conference Program Committee reviewed 54 submissions for the conference and eight submissions for the PhD workshop. For the PhD workshop seven papers were accepted and one was rejected. For the conference 24 submissions were accepted as full papers, 18 as short papers, five as posters, whereas 12 submissions were rejected. According to the conference program, these 47 oral presentations were structured into 14 sessions including Internet of Things and Cognitive Systems, Data Integration and Data Analysis, Knowledge Representation, Ontologies and Applications, Data Analysis and Applications, Conceptual and Data Models, Advanced Data Analysis Methods, Data Analysis in Astronomy, Text Search and Processing, Distributed Computing and Applications of Machine Learning, Research Data Infrastructures, and Information Extraction from Text.

Although most of the presentations were dedicated to the results of projects conducted in the research organizations based in the Russian Federation including Chelyabinsk, Ekaterinburg, Kazan, Moscow, Novosibirsk, Obninsk, St. Petersburg, Tomsk, Tver, Ulyanovsk, the conference has acquired features of internationalization. This move is witnessed by 13 talks (six of them were invited) prepared by well-known foreign researchers from such countries as Armenia (Yerevan), Belgium (Brussels), Germany (Munich, Hamburg, Kiel), Great Britain (Harwell), Hungary (Budapest), Finland (Espoo, Helsinki), France (Paris), Kazakhstan (Almaty), The Netherlands (Leiden).

For the proceedings, 12 papers were selected by the Program Committee (nine peer reviewed and three invited papers) and after careful editing they formed the content of the post-conference volume structured into seven sections including FAIR Data Infrastructures, Interoperability and Reuse (three papers), Knowledge Representation (one paper), Data Models (one paper), Data Analysis in Astronomy (one paper), Text Search and Processing (two papers), Distributed Computing (one paper), Information Extraction from Text (three papers).

The chairs of Program Committee express their gratitude to the Program Committee members for carrying out the reviewing of the submissions and selection of the papers for presentation, to the authors of the submissions, as well as to the Russian Foundation for Basic Research and the Fund "League online media" for the financial support to the conference. The Program Committee of the conference appreciates the possibility to use the Conference Management Toolkit (CMT) sponsored by Microsoft Research, which provided great support during various phases of the paper submission and reviewing process. Finally, we thank Springer for publishing this proceedings volume, containing the revised invited and selected research papers, in their CCIS series.

April 2019   Yannis Manolopoulos
                                                             Sergey Stupnikov

Organization

General Chair

Igor Sokolov — Federal Research Center Computer Science and Control of RAS, Russia

Program Committee Co-chairs

Leonid Kalinichenko — Federal Research Center Computer Science and Control of RAS, Russia

Yannis Manolopoulos — Aristotle University of Thessaloniki, Greece

FAIR Workshop Co-chairs

Peter Wittenburg — Max Planck Computing and Data Facility, Garching/Munich, Germany

Leonid Kalinichenko — Federal Research Center Computer Science and Control of RAS, Russia

PhD Workshop Chair

Roman Samarev — Bauman Moscow State Technical University, Russia

Organizing Committee Co-chairs

Vladimir Sukhomin — Lomonosov Moscow State University, Russia

Victor Zakharov — Federal Research Center Computer Science and Control of RAS, Russia

Organizing Committee

Elena Zubareva — Lomonosov Moscow State University, Russia

Dmitry Briukhov — Federal Research Center Computer Science and Control of RAS, Russia

Nikolay Skvortsov — Federal Research Center Computer Science and Control of RAS, Russia

Yulia Trusova — Federal Research Center Computer Science and Control of RAS, Russia

Supporters

Russian Foundation for Basic Research
Federal Research Center Computer Science and Control of the Russian Academy
of Sciences (FRC CSC RAS)
Moscow ACM SIGMOD Chapter

Coordinating Committee

Igor Sokolov (Co-chair)	Federal Research Center Computer Science and Control of RAS, Russia
Nikolay Kolchanov (Co-chair)	Institute of Cytology and Genetics, SB RAS, Novosibirsk, Russia
Leonid Kalinichenko (Deputy chair)	Federal Research Center Computer Science and Control of RAS, Russia
Arkady Avramenko	Pushchino Radio Astronomy Observatory, RAS, Russia
Pavel Braslavsky	Ural Federal University, SKB Kontur, Russia
Vasily Bunakov	Science and Technology Facilities Council, Harwell, Oxfordshire, UK
Alexander Elizarov	Kazan (Volga Region) Federal University, Russia
Alexander Fazliev	Institute of Atmospheric Optics, RAS, Siberian Branch, Russia
Alexei Klimentov	Brookhaven National Laboratory, USA
Mikhail Kogalovsky	Market Economy Institute, RAS, Russia
Vladimir Korenkov	JINR, Dubna, Russia
Mikhail Kuzminski	Institute of Organic Chemistry, RAS, Russia
Sergey Kuznetsov	Institute for System Programming, RAS, Russia
Vladimir Litvine	Evogh Inc., California, USA
Archil Maysuradze	Moscow State University, Russia
Oleg Malkov	Institute of Astronomy, RAS, Russia
Alexander Marchuk	Institute of Informatics Systems, RAS, Siberian Branch, Russia
Igor Nekrestjanov	Verizon Corporation, USA
Boris Novikov	St.-Petersburg State University, Russia
Nikolay Podkolodny	ICaG, SB RAS, Novosibirsk, Russia
Aleksey Pozanenko	Space Research Institute, RAS, Russia
Vladimir Serebryakov	Computing Center of RAS, Russia
Yury Smetanin	Russian Foundation for Basic Research, Moscow
Vladimir Smirnov	Yaroslavl State University, Russia
Sergey Stupnikov	Federal Research Center Computer Science and Control of RAS
Konstantin Vorontsov	Moscow State University, Russia
Viacheslav Wolfengagen	National Research Nuclear University MEPhI, Russia
Victor Zakharov	Federal Research Center Computer Science and Control of RAS, Russia

Program Committee

Alexander Afanasyev	Institute for Information Transmission Problems, RAS, Russia
Plamen Angelov	Lancaster University, UK
Arkady Avramenko	Pushchino Observatory, Russia
Ladjel Bellatreche	Laboratory of Computer Science and Automatic Control for Systems, National Engineering School for Mechanics and Aerotechnics, Poitiers, France
Pavel Braslavski	Ural Federal University, Yekaterinburg, Russia
Vasily Bunakov	Science and Technology Facilities Council, Harwell, UK
Evgeny Burnaev	Skolkovo Institute of Science and Technology, Russia
George Chernishev	Saint-Petersburg State University, Russia
Yuri Demchenko	University of Amsterdam, The Netherlands
Boris Dobrov	Research Computing Center of MSU, Russia
Alexander Elizarov	Kazan Federal University, Russia
Alexander Fazliev	Institute of Atmospheric Optics, SB RAS, Russia
Yuriy Gapanyuk	Bauman Moscow State Technical University, Russia
Vladimir Golenkov	Belarusian State University of Informatics and Radioelectronics, Belarus
Vladimir Golovko	Brest State Technical University, Belarus
Olga Gorchinskaya	FORS Group, Moscow, Russia
Evgeny Gordov	Institute of Monitoring of Climatic and Ecological Systems, SB RAS, Russia
Valeriya Gribova	Institute of Automation and Control Processes, FEB RAS, Far Eastern Federal University, Russia
Maxim Gubin	Google Inc., USA
Natalia Guliakina	Belarusian State University of Informatics and Radioelectronics, Belarus
Ralf Hofestadt	University of Bielefeld, Germany
Abdulrahman Kaitoua	German Research Center for Artificial Intelligence, Germany
George Karypis	University of Minnesota, Minneapolis, USA
Nadezhda Kiselyova	IMET RAS, Russia
Alexei Klimentov	Brookhaven National Laboratory, USA
Mikhail Kogalovsky	Market Economy Institute, RAS, Russia
Vladimir Korenkov	Joint Institute for Nuclear Research, Dubna, Russia
Sergey Kuznetsov	Institute for System Programming, RAS, Russia
Dmitry Lande	Institute for Information Recording, NASU, Ukraine
Vladimir Litvine	Evogh Inc., California, USA
Giuseppe Longo	University of Naples Federico II, Italy
Natalia Loukachevitch	Lomonosov Moscow State University, Russia
Ivan Lukovic	University of Novi Sad, Serbia
Oleg Malkov	Institute of Astronomy, RAS, Russia
Archil Maysuradze	Lomonosov Moscow State University, Russia

Manuel Mazzara	Innopolis University, Russia
Alexey Mitsyuk	National Research University Higher School of Economics, Russia
Xenia Naidenova	S. M. Kirov Military Medical Academy, Russia
Dmitry Namiot	Lomonosov Moscow State University, Russia
Igor Nekrestyanov	Verizon Corporation, USA
Gennady Ososkov	Joint Institute for Nuclear Research, Russia
Panos Pardalos	Department of Industrial and Systems Engineering, University of Florida, USA
Nikolay Podkolodny	Institute of Cytology and Genetics SB RAS, Russia
Jaroslav Pokorny	Charles University in Prague, Czech Republic
Natalia Ponomareva	Scientific Center of Neurology of RAMS, Russia
Alexey Pozanenko	Space Research Institute, RAS, Russia
Tilmann Rabl	Technische Universität Berlin, Germany
Andreas Rauber	Vienna Technical University, Austria
Roman Samarev	Bauman Moscow State Technical University, Russia
Timos Sellis	Swinburne University of Technology, Australia
Vladimir Serebryakov	Computing Centre of RAS, Russia
Nikolay Skvortsov	Federal Research Center Computer Science and Control of RAS, Russia, Russia
Vladimir Smirnov	Yaroslavl State University, Russia
Leonid Sokolinskiy	South Ural State University, Russia
Valery Sokolov	Yaroslavl State University, Russia
Sergey Stupnikov	Federal Research Center Computer Science and Control of RAS, Russia, Russia
Alexander Sychev	Voronezh State University, Russia
Bernhard Thalheim	University of Kiel, Germany
Theodoros Tzouramanis	University of the Aegean, Greece
Alexey Ushakov	University of California, Santa Barbara, USA
Dmitry Ustalov	University of Mannheim, Germany
Natalia Vassilieva	Hewlett-Packard, Russia
Pavel Velikhov	Finstar Financial Group, Russia
Alexey Vovchenko	Federal Research Center Computer Science and Control of RAS, Russia, Moscow, Russia
Peter Wittenburg	Max Planck Computing and Data Facility, Garching/Munich, Germany
Vladimir Zadorozhny	University of Pittsburgh, USA
Yury Zagorulko	Institute of Informatics Systems, SB RAS, Russia
Victor Zakharov	Federal Research Center Computer Science and Control of RAS, Russia, Russia
Oleg Zhizhimov	Institute of computing technologies of SB RAS, Russia
Sergey Znamensky	Institute of Program Systems, RAS, Russia

Contents

FAIR Data Infrastructures, Interoperability and Reuse

FAIR Principles and Digital Objects: Accelerating Convergence on a Data Infrastructure

Erik Schultes[1]($\boxtimes$) and Peter Wittenburg[2]($\boxtimes$)

[1] GO FAIR International Support and Coordination Office, Poortgebouw N-01, Rijnsburgerweg 10, 2333 AA Leiden, The Netherlands
erik.schultes@go-fair.org
[2] Max Planck Computing and Data Facility, Gemeindeweg 55, 47533 Kleve, Germany
peter.wittenburg@mpi.nl

Abstract. As Moore's Law and associated technical advances continue to bulldoze their way through society, both exciting possibilities and severe challenges emerge. The upside is the explosive growth of data and compute resources that promise revolutionary modes of discovery and innovation not only within traditional knowledge disciplines, but especially between them. The challenge, however, is to build the large-scale, widely accessible, persistent and automated infrastructures that will be necessary for navigating and managing the unprecedented complexity of exponentially increasing quantities of distributed and heterogenous data. This will require innovations in both the technical and social domains. Inspired by the successful development of the Internet and leveraging the Digital Object Framework and FAIR Principles (for making data Findable, Accessible, Interoperable and Reusable by machines) the GO FAIR initiative works with voluntary stakeholders to accelerate convergence on minimal standards and working implementations leading to an Internet of FAIR Data and Services (IFDS). In close collaboration with GO FAIR and DONA, the RDA GEDE and C2CAMP initiatives will continue its FAIR DO implementation efforts..

Keywords: Data analytics · Data management · Data intensive science · Digital libraries · Digital objects · FAIR data

1 Introduction

Existing data stewardship practices are highly inefficient. Numerous studies indicate that data scientists both in academia and industry spend 70–80% of their time on mundane, manual procedures to locate, access, and format data for reuse [1, 2]. Methodological legacies inherited from a pre-digital era (e.g., poor capture of metadata, broken links to various research assets) and outdated professional incentives (e.g., only rewarding publication of research articles rather than also datasets and other research outputs) contribute to massive data loss and a well-documented reproducibility crisis [3–5]. Coupled with the exponential increases in data volumes (driven by, among other things, high through-put instrumentation and IoT data streams) the urgency for

© Springer Nature Switzerland AG 2019
Y. Manolopoulos and S. Stupnikov (Eds.): DAMDID/RCDL 2018, CCIS 1003, pp. 3–16, 2019.
https://doi.org/10.1007/978-3-030-23584-0_1

automated, commonly usable and persistent data infrastructures (i.e., a datanet for Machines) is increasingly recognised by numerous national and international organisations, science funders and industry [6–11]. Despite the urgent need, building a generalised, ubiquitous, data infrastructure that is widely used by diverse stakeholders is an inherently distributed and difficult process to direct. Knowing this to be the case, members of several RDA groups started the C2CAMP initiative [12] to join results and to build a testbed for a Digital Objects based infrastructure which will help overcoming the huge inefficiencies in data intensive science. In parallel, the GO FAIR initiative was launched to also accelerate data infrastructure development by leveraging general patterns of phased development described in other revolutionary infrastructures, including the Internet and the World Wide Web (WWW) [13].

2   Learning from Previous Revolutionary Infrastructures

Revolutionary Infrastructures (for example, transportation, electrification, telecommunications, and computer networks) follow five phases of development [14, 15]: (1) Vision: New discoveries and technologies lead to the anticipation of broad new application spaces; (2) Creolization: Inspired by the Vision, numerous experimental implementations are created, resulting in an uneven landscape of independently developed prototypes; (3) Attraction: Some solutions prove more viable, and are effectively generalised to achieve a simplified set of 'universal principles' that attract the attention of others working in the field; (4) Convergence: Various Attractors voluntarily decide to bridge otherwise isolated application solutions, and a compelling global infrastructure begins to emerge at the expense of the many other possibilities; (5) Exploitation: As widespread commitment to a particular implementation emerges, economy of scale kicks in, and what was hard and cost-prohibitive, now becomes easy and affordable. Users in the Exploitation phase might not even be aware of the infrastructure systems they routinely use (e.g., most users of the internet are blissfully ignorant of TCP/IP).

In the specific case of the Internet, there had been early *Visions* of interlinked computers throughout the 1950s and 1960s. By 1969, ARPAnet had initiated the phases of *Creolization* (and later *Attraction*) with the co-existence of multiple, specialised solutions, e.g., X25, Ethernet, ARCNET, and others. This work demonstrated the feasibility of computer networks and drew the attention of large investors (e.g., IBM, DEC). But this investment resulted in numerous incompatible standards that first drove insights but later slowed progress. *Convergence* was eventually triggered with TCP/IP protocols (early 1970s) and the 7-layer ISO/OSI reference model (early 1980s). This was because, in particular, the minimal TCP/IP standard allowed various networks to interoperate while at the same time maintaining maximum freedom to engineer solutions at the implementation layer 'below' and application layer 'above' (creating the so-called "hourglass" architecture of the Internet, with TCP/IP at the narrow waist). It was working implementations (however embryonic) and the simplicity of the hourglass approach that motivated influential decision makers "to move towards using TCP/IP as universal for implementing global computer networking". With a stabilized universal in place, *Exploitation* soon followed, with rapid investment in both hardware and software, that is the now familiar story of the Internet. By 1992, the Internet Society was set up to coordinate further develop TCP/IP approaches to networking.

It is important to note that the use of TCP/IP has always been voluntary, and at no time was its use ever required. Indeed, top-down enforcement policies would likely have killed its effectiveness as an attractor. Instead, once a 'critical mass' of influential users had adopted TCP/IP, the larger community followed, driving convergence. An analogous pattern of development (voluntary use, attractor effect in the community) occurred soon after with the formation of the WWW, in this case with HTTP playing the role of TCP/IP. The significance of this historical insight can not be understated. It enables some degree of coordination in the development of new infrastructures, because only a relatively few (albeit influential) users need be convinced to invest in a particular technology. Once the 'critical mass' is assembled, the 'long tail' of community stakeholders will likely follow.

Even before the 2000's, visionaries had already anticipated the need for a general-purpose data infrastructure. Digital Object based infrastructures such as the Digital Object Architecture [16], systems supporting Persistent Identifiers (PIDs) and the Semantic Web (a framework for knowledge representation built on top of existing Internet and WWW infrastructures) appeared as an important component, ensuring both data interoperation and machine readability. Since then, difficult problems in this space have been investigated resulting in a plenum of new, co-existing methods, languages, software and specialised hardware, producing by now, a protracted period of Creolization. By 2012 the Attraction phase was underway with public discussions about component specifications, principles and procedures for semantically enabled data infrastructures [17, 18]. RDA was officially started in 2013 as a broad group of data experts including now more than 7000 persons from more than 120 countries and had first results from working groups in 2014. Some of the RDA experts recognised the need to bring the various results together and started first the RDA Data Fabric group [19] to identify Digital Objects as the common ground and to specify additional needs. Then, emerging from RDA, the C2CAMP collaboration was created to not only specify procedures and interfaces, but to start working on a joint testbed in close collaboration with the DONA foundation [20]. Later the GEDE collaboration adopted the DO topic and subsequently organising more than 150 data experts from about 47 European research infrastructures to participate in the discussions on Digital Objects.

By early 2014, in a workshop hosted by the Lorentz Center (Leiden), the above mentioned discussion culminated in the generalised and broadly applicable FAIR Principles for data reuse [21, 22]. In a now widely cited commentary (indicative of the Attraction phase) [23], the FAIR approach had been defined as "Data and services that are findable, accessible, interoperable, and re-usable both for machines and for people" and 15 high-level Principles had been articulated, Fig. 1.

Findable:

F1 (meta)data are assigned a globally unique and persistent identifier;

F2 data are described with rich metadata;

F3 metadata clearly and explicitly include the identifier of the data it describes;

F4 (meta)data are registered or indexed in a searchable resource;

Interoperable:

I1 (meta)data use a formal, accessible, shared, and broadly applicable language for knowledge representation.

I2 (meta)data use vocabularies that follow FAIR principles;

I3 (meta)data include qualified references to other (meta)data;

Accessible:

A1 (meta)data are retrievable by their identifier using a standardized communications protocol;

A1.1 the protocol is open, free, and universally implementable;

A1.2 the protocol allows for an authentication and authorisation procedure, where necessary;

A2 metadata are accessible, even when the data are no longer available;

Reusable:

R1 meta(data) are richly described with a plurality of accurate and relevant attributes;

R1.1 (meta)data are released with a clear and accessible data usage license;

R1.2 (meta)data are associated with detailed provenance;

R1.3 (meta)data meet domain-relevant community standards;

Fig. 1. The 15 FAIR principles ensuring machine findability, accessibility, interoperation and re-use of digital resources.

Immediately following their publication (April 2016), the FAIR Principles (and later, the corresponding FAIR Metrics [24] and FAIR Maturity Indicators [25]) have been acting as a powerful attractor in the emerging data infrastructure.

Following the previous examples, the Convergence phase of the data infrastructure will commence once a 'critical mass' of users commits to particular, minimal specification for automatic routing of FAIR data and services.

In the meantime the strong relationship between the FAIR Principles and FAIR Digital Objects has been observed by the GO FAIR and C2CAMP/GEDE experts [26]. These groups are now working together to harmonise the DO and FAIR approaches into a formally defined "FAIR DO", with the aim to accelerate convergence on a globally distributed data infrastructure. A data infrastructure will likely be substantially more complex than its predecessors in that a FAIR Digital Objects based Internet of FAIR Data and Services (IFDS) necessitates the wide acceptance of the DO Interface Protocol, the use of the potential of the globally available Handle System to solve the binding challenge and to elaborate on semantically enabled metadata descriptions. The 'FAIRification' of digital resources is not trivial, and widespread application will require an ecosystem of methods, tooling, services and training that help communities of diverse stakeholders to create and use FAIR resources. While C2CAMP/GEDE and DONA will showcase a stable DO-based eco-system of infrastructures, GO FAIR will support and coordinate bottom-up community initiatives that aim to 'Make FAIR easy" [27].

3 FAIR Digital Objects

3.1 Digital Objects

Digital Objects were already introduced in an early paper by Kahn & Wilensky in 1995 and then in an updated version in 2006 [28, 29]. As Wittenburg et al. [30] have shown the concept is very much related with computer science concepts such as "object-oriented programming" [31], "abstract data types" [32] and "object stores" [33] which are at the basis of state-of-the-art cloud systems such as Amazon's S3. We can therefore claim that the concept of "objects", is closely related with ideas such as "encapsulation", "virtualization", and "interfacing by defined methods", has shown its great importance to help designing complex systems.

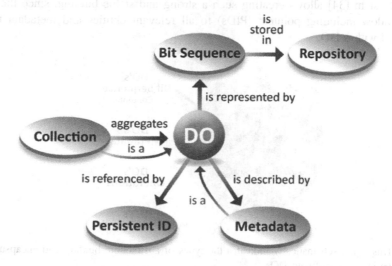

Fig. 2. This figure indicates the core data model as it was worked out within the RDA group Data Foundation and Terminology (DFT).

In 2014, the RDA group "Data Foundation and Terminology" (DFT) published its results on a core data model and the corresponding basic terminology. It summarized the discussions about Digital Objects (DO) as (see Fig. 2):

- DOs are at the core of a proper data organizations in so far as it has the capacity to bind crucial entities which are necessary for a stable and reusable domain of data;
- DOs have a bit sequence (content) which can be stored in various repositories, are referenced by a unique and persistent identifier (PID) issued by a trustworthy globally available resolution system and is described by various types of metadata that can include descriptive, system, access rights, license, contractual, transactional and other kinds of meta information about the DO;
- Metadata itself are DOs;

- DOs can be combined to collections which also are DOs, i.e. have a PID and are described by metadata;
- DOs can include all kinds of digital information such as data, software, configurations, representations of persons, institutions, semantic concepts, etc.

We can also look schematically at DOs from a different point of view, if we extend the above definition by encapsulation principles as being introduced by the RDA group "Data Type Registry". One of the metadata types describing a DO is its "type" which is summarizing several technical metadata attributes. A Data Type Registry allows users to relate data types with operations which are also DOs of a specific type. These defined operations allow users to realize the encapsulation principle as requested by Abstract Data Types. Figure 3 indicates this encapsulation which can be implemented when strong and stable binding is being realized. The usage of PID systems such as the Handle System [34] allows creating such a strong and stable binding, since the PID records allow including pointers (PIDs) to all relevant entities and metadata types associated with a DO.

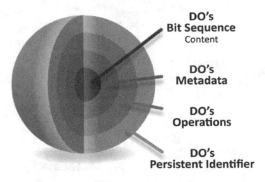

Fig. 3. This figure schematically indicates the types of abstraction, binding and encapsulation that can be implemented with DOs.

Recently, a second version of a protocol to interact with DOs, the DO Interface Protocol (DOIPV2.0), has been opened for broad discussion by DONA. It basically describes how clients interact with DOs where all involved actors are represented by PIDs. DOIP is meant to have a relevance that is comparable to TCP/IP for the Internet, i.e. it should become a fundamental protocol to manage and exchange digital objects.

The definition of the term "Digital Object" in the DOIP document is intended to be restricted in its focus on the minimalistic and operational nature of the protocol, i.e. a DO has a bit sequence, a PID and a type. Although elegant in its simplicity, this minimal definition itself gives no specification for the recording the scientific semantics or other domain knowledge that is equally important in routing and processing research data and services in general included in metadata. Recently, the RDA DFT group started addressing this issue by augmenting the minimal DO definition in the context of the FAIR Principles itself, defining the "FAIR Digital Object", since it includes the strong binding of different types of metadata which are important for the interpretation and access and reuse of the bit sequences.

The C2CAMP initiative is devoted to implement a FAIR DO based infrastructure including understanding DOs as active entities that have methods associated with them. A broad discussion started in Europe in the realm of GEDE [35] involving 150 experts from about 50 research infrastructures to intensify the discussions not only about the potential of FAIR DOs to build federative data infrastructures, but in particular to also use FAIR DOs to systematically structure the domain of digital entities in scientific disciplines. A recent workshop [36] combining these two roles of FAIR DOs showed globally organised research communities such as biodiversity, climate modeling and language research have far going plans to use their potential to increase trust, to define clear anchors for a complex system of annotation layers, to better utilize automatic workflow frameworks and much more. Moving forward, there is now increasing interest in the fusion of DO and FAIR approaches both at the conceptual and technical levels.

3.2   FAIR Principles

The original publication announcing the FAIR Principles does not discuss implementation choices. Given that many different combinations of technology choices and use of standards could conceivably implement the FAIR Principles, the GO FAIR initiative was launched in late 2017 by the Dutch, German and French governments as a means to pragmatically accelerate community Convergence. The initial vehicle for GO FAIR is the International Support and Coordination Office (GFISCO). Following the examples of the Internet and WWW, the GFISCO operates through voluntary stakeholder participation attempting to reach a 'critical mass' of users committed to a set of absolute minimal technology specifications. Beyond these minimal specifications, there is unrestricted room to innovate.

GFISCO is stakeholder governed, and includes researchers from specialized knowledge domains (e.g., earth sciences [37], chemistry [38]) but also policy bodies (e.g., CODATA, RDA, FORCE11), publishers (e.g., Elsevier, Springer-Nature), repositories (e.g., Figshare), and funding agencies (e.g., The American NSF and NIH, the Health Research Board of Ireland, and the Dutch ZonMW). GFISCO brokers among stakeholders, the choice of standards implementing the functions of the FAIR Principles and emerging best practices leading to the Internet of FAIR Data and Services. GFISCO operates via supporting and coordinating Implementation Networks (INs), which are voluntary international consortia that self-organize (and are self-funded) to implement elements of the IFDS. GO FAIR INs belong to 3 broad topical pillars: GO BUILD, GO TRAIN and GO CHANGE.

GO BUILD focuses on the technological aspects of the IFDS, including the design and building of reference implementations for elements composing the IFDS such as FAIR Metrics, FAIR Data Points [39, 40], FAIRification tools and other FAIR-compliant services. Currently, there are 8 INs under the GO BUILD pillar.

Other technology-related activities in GO FAIR include ongoing "Metadata for Machines" workshops and "Community Challenges", aiming to help communities achieve adoption of globally unique and persistent identifiers, agree on common metadata representation formats, agree on a minimal set of generic metadata content and define domain-relevant community standards.

The overall objective of the GO TRAIN pillar is to create a scalable framework that is used in higher education programs and throughout industry to train large numbers of certified data stewards (estimated to be 500,000 for Europe [41], millions more worldwide). GO TRAIN supports and coordinates two activities: (1) The development of canonical training curricula focused on FAIR Data Stewardship; (2) The development of certification schema for competencies in FAIR Data Stewardship (providing professional career trajectories, which in turn, are intended to drive rapid uptake of FAIR practices among diverse stakeholders). Currently there are two GO TRAIN INs. The first is the Training Frameworks IN which aims to develop schema for FAIR Data Stewardship education (including train-the-trainer curricula and endorsement specifications), with lenses for Managers, Principal Investigators and Data Stewards themselves. Secondly, The FAIR Curriculum IN will re-use the Carpentries Open, community based curriculum development model [42] to develop novel modular lessons for FAIR data stewardship.

The overall purpose of the GO CHANGE pillar to support and coordinate systemic culture change that transforms existing data management practices into the respected profession of data stewardship. This includes the development of new funding schema, sustainability strategies, and business models. GO CHANGE stakeholders range from international policy makers and national governments to organisation managers and front-line data producers and data stewards. A key IN for GO CHANGE is a FAIR resource hub that aggregates multiple resources for FAIR data stewardship planning, compliance, and assessment.

4   GO FAIR, DO FAIR

A preliminary analysis can easily show that there is a close but highly complementary relationship between the FAIR Principles and the concept of FAIR Digital Objects.

4.1   Data to be Findable

The DO model is widely compliant with the F-dimension of the FAIR principles and gives an implementation mechanism. The DO model is explicit in how to do things - in particular the binding of different informational entities associated with the DO to guarantee FAIRness - but does not specify the possible usage of DO's content. It includes certified repositories as active components and care takers of data and does not make statements about the content of metadata, since this is very much purpose dependent and domain specific. Whereas DOs are agnostic about its content and treat all kinds of content (data, metadata, software, semantic assertions, etc.) the same way the FAIR principle F2 requests rich metadata for findability that can be both generic and domain focused.

4.2   Data to be Accessible

Entirely consistent with the Accession-Related FAIR Principles, the DO Core Model enables the building of infrastructure that makes data and metadata accessible since it supports all requirements with respect to open and free to use protocols but also proper authentication and authorization where necessary. Except for the PID infrastructure which is an essential element of DO based infrastructures, the DO model assigns the responsibility to repositories to define policies and implement appropriate mechanisms. As such, authentication and authorization aspects need to be taken care by the inter-acting distributed components on the Internet of FAIR Data and Services. The FAIR Principle A2 stipulates a condition that is only implicit in the DO model, which is that metadata should persist, even if the original data are deleted or in some way no longer available.

4.3   Data to be Interoperable

DOs take care of interoperability at the level of data organisation due to its inherent binding concept and its stable linking based on specific PIDs such as Handles and this in a way that is machine actionable. The DO Interface Protocol is a universal mech-anism to interact with DOs independent of how repositories organise and model their digital entities. Although with respect to other interoperability layers such as structural and semantical encoding of content the DO concept is agonistic, it does facilitate the operational work at these levels by allowing users to use the DO model for all kinds of digital entities and thus guaranteeing stable binding that is necessary for interoperation. However, again we see the complementarity between the FAIR Principles and the DO model, in that the 3 Interoperation-related FAIR Principles are explicit about rich, qualified semantic encoding.

4.4   Data to be Reusable

The DFT Core Model explicitly mentions the role of key properties of DO's content being part of the PID record or being referred to by stable and persistent links. Due to strong typing as suggested by the RDA Kernel Information group of all these attributes machine actionability is given a great advantage. The binding concept of the DFT Core Model enables the linking of various aspects closely related with the DO such as provenance, smart contracts (actionable licenses), transaction records and even more that go beyond the FAIR principles. The DO Core Model is agnostic with respect to the concrete specifications, since it respects (indeed, expects) that other groups such as W3C (PROV), the blockchain community, etc. are providing mechanisms and defi-nitions which will be used to implement special wishes. Thus again we can say that the DO concept facilitates the implementation of the FAIR requirements, although the FAIR components have the capability and mandate to express rich and nuanced semantics.

4.5 Summary

Due to their complementarity we see GO FAIR and DO activities as a giant step towards improving data practices and it was a logical step for the C2CAMP/GEDE initiatives to become an Implementation Network in GO FAIR and to also align discussions with the GEDE DO Topic Group as well. GO FAIR distinguishes three major areas of work (see Fig. 4) to build FAIR compliant infrastructures: data, tools and compute resources, which in the DO domain are Digital Objects of different types. All three areas share one central infrastructure, the turbine's driving axis. To expand this metaphor one could imagine the DOs to be the driving axis that combines all three areas and the DO Interface Protocol and the protocol to resolve persistent identifiers as the underlying basic protocols all areas are using. While DOs implement the F and A dimensions of FAIR more or less directly, they facilitate the I and R dimensions.

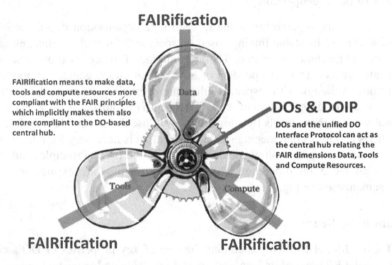

FAIRification

FAIRification means to make data, tools and compute resources more compliant with the FAIR principles which implicitly makes them also more compliant to the DO-based central hub.

Data

DOs & DOIP

DOs and the unified DO Interface Protocol can act as the central hub relating the FAIR dimensions Data, Tools and Compute Resources.

Tools   Compute

FAIRification   **FAIRification**

Fig. 4. This figure indicates the major dimensions of the GO FAIR work and the interpretation of Digital Objects being the driving wheel combining all three dimensions.

The FAIR Digital Object approach provides technical solutions needed to implement FAIR principles. In particular, federated systems such as intended, for example, by the European Open Science Cloud will need such a basic interoperability layer to achieve the required scalability, stability and FAIRness. Building such a comprehensive and expensive infrastructure eco-system will need to be based on solid fundaments as offered by the FAIR principles and FAIR Digital Objects to overcome major hurdles in making data more reusable.

5 Participating

5.1 RDA GEDE DO Topic Group

GEDE, the Group of European Data Experts, is organised within RDA and defines so-called topic groups to allow interested experts to work on specific thematic topics. One of these topics are the FAIR Digital Objects where 150 distinguished data experts from about 50 European research infrastructures and some international colleagues are discussing intensively about how to improve data work by adopting FAIR DOs. Currently, a set of more than 30 use cases has been presented by different communities which will lead to a new paper on FAIR DOs driven by scientific interests. Participation in GEDE DO is open to anyone interested.

5.2 C2CAMP

C2CAMP is a global collaboration of experts who want to build DO-based infrastructures and tools that emerged from the work in RDA groups and that closely collaborates with the GEDE DO topic group. C2CAMP participation is open for anyone who wants to actively contribute to the FAIR DO testbed.

In 2018 C2CAMP joined GO FAIR as an implementation network to foster the interaction with other implementation networks.

5.3 GO FAIR Implementation Network

GO FAIR INs foster a collaborative community of harmonized practice which leads to Convergence and allows members to 'speak with one voice' on critical issues regarding FAIR data infrastructures. Anyone (i.e., a person, an institution or a network organisation) can join an existing or create a new GO FAIR IN [43]. The list of current GO FAIR INs can be found at the GO FAIR website [44]. The requirements to become an IN are minimal: (1) have a plan to implement an element of the IFDS (including adequate resourcing to accomplish the proposed goals); (2) comply with the GO FAIR Rules of Engagement (essentially, commitment to the FAIR Principles and 'no vendor lock-in[1]'); (3) have sufficient critical mass to be regarded as thought leaders in the field of expertise.

6 Conclusions

As described by Wittenburg and Strawn [14] we see trends to convergence finding in the data domain. Two major action lines have been kicked off almost in parallel: on the one hand by the RDA groups that worked together in the RDA Data Fabric group and later started the C2CAMP and GEDE collaboration; on the other hand by the group working on the FAIR principles and provided the background in which the GO FAIR initiative was launched. Both initiatives saw the need to turn specifications into active

---

[1] https://www.go-fair.org/implementation-networks/rules-of-engagement/.

implementation work and thus contributing to the emerging practical eco-system of data infrastructures. In addition, they understood that FAIR principles and FAIR Digital Objects are complementary.

A new wave of investments in large research and data infrastructures can be observed including the European Open Science Cloud and national science clouds in most of the European member states. The relevant actors sense that what they are aiming at is finally a complex enterprise with many open questions - their undertaking is a huge experiment that will lead to a transformation of science. Two of these open questions are: how complexity can be broken down and how a stable fundament for the coming decades can be achieved that will not hamper the needed progress in science. It should be noted that the severity of obstacles to data reuse is driven ultimately by Big Data (Moore's Law) and in this sense, the problems extend far beyond the research domain. Industry is confronted with similar challenges and thus may need to find similar solutions if it will completely be locked in proprietary platforms.

We recommend therefore following the trend to FAIR data and doing this by implementing the FAIR DO concept that has as core elements globally resolved persistent identifiers and the Digital Object Interface Protocol - all being open specifications governed by the non-profit Swiss DONA Foundation.

Acknowledgments. We thank the many collaborators in C2CAMP, RDA GEDE and GO FAIR to contribute to the ongoing discussions which led to this publication.

References

1. Stehouwer, H., Wittenburg, P.: RDA Europe: Data Practices Analysis (2018). http://hdl. handle.net/11304/6e1424cc-8927-11e4-ac7e-860aa0063d1f
2. Data Scientist Report. Crowdflower (2017). https://visit.crowdflower.com/WC-2017-Data-Science-Report_LP.html
3. Schloss, P.D.: Identifying and overcoming threats to reproducibility, replicability, robustness, and generalizability in microbiome research. mBio **9**(3), e00525–18 (2018). https://doi.org/10.1128/mBio.00525-18
4. Gorgolewski, K.J., Poldrack, R.A.: A practical guide for improving transparency and reproducibility in neuroimaging research. PLoS Biol. **14**(7), e1002506 (2016). https://doi.org/10.1371/journal.pbio.1002506
5. Mons, D.: Data Stewardship for Open Science: Implementing FAIR Principles, 1st edn. Chapman and Hall, Boca Raton (2018)
6. Research Data Alliance. https://www.rd-alliance.org
7. Implementation Roadmap for the European Open Science Cloud (2018) http://www.esfri.eu/ri-world-news/implementation-roadmap-european-open-science-cloud
8. New Models of Data Stewardship. NIH Data Commons. https://commonfund.nih.gov/commons
9. How expensive is FAIR compliance and how expensive is it to not be FAIR compliant. RDA 11th Plenary BoF meeting (2018). https://rd-alliance.org/how-expensive-fair-compliance-and-how-expensive-it-not-be-fair-compliant-rda-11th-plenary-bof
10. G7 Science Ministers' Communique (2017). http://www.g7.utoronto.ca/science/2017-G7-Science-Communique.pdf

11. Progress Towards the European Open Science Cloud: GO FAIR Office Established, Global ActionPlatform (2017). http://globalactionplatform.org/post/progress-towards-the-european-open-science-cloud-go-fair-office-established
12. Lannom, L.: Managing digital objects in an expanding science ecosystem (2017). https://www.rd-alliance.org/sites/default/files/CENDI-15.Nov_.17-Lannom-Final-2.pdf
13. Hughes, T.P.: Networks of Power: Electrification in Western Society 1880–1930. Johns Hopkins University Press, Baltimore (1983)
14. Wittenburg, P., Strawn, G.: Common Patterns in Revolutionary Infrastructures and Data. US National Academy of Sciences (2018). https://www.rd-alliance.org/sites/default/files/Common_Patterns_in_Revolutionising_Infrastructures-final.pdf
15. International DAITF Workshop at the ICRI 2012 Conference (2012). http://www.icri2012.dk/www.ereg.me/ehome/index06e1.html
16. Digital Object Architecture. https://www.dona.net/digitalobjectarchitecture
17. Berg-Cross, G., Ritz, R., Wittenburg, P.: Research data alliance, data foundation & terminology group core terms and model (2016). http://hdl.handle.net/11304/5d760a3e-991d-11e5-9bb4-2b0aad496318
18. The FAIR Data Principles. FORCE11 (2016). https://www.force11.org/group/fairgroup/fairprinciples
19. RDA Data Fabric. https://www.rd-alliance.org/group/data-fabric-ig.html
20. DONA Foundation. https://www.dona.net/
21. Jointly designing a data FAIRPORT. Lorentz Center faculty of Science of Leiden University, Leiden The Netherlands (2014). https://www.lorentzcenter.nl/lc/web/2014/602/info.php3?wsid=602
22. FAIR Principles Explained. GO FAIR. https://www.go-fair.org/fair-principles/
23. Wilkinson, M.D., et al.: The FAIR Guiding Principles for scientific data management and stewardship. Sci. Data 3 (2016). https://doi.org/10.1038/sdata.2016.18
24. Wilkinson, M.D., et al.: A design framework and exemplar metrics for FAIRness. Sci. Data 5, 180118 (2018). https://doi.org/10.1038/sdata.2018.118
25. Wilkinson, M.D., et al.: Evaluating FAIR-compliance through an objective, automated, community-governed framework. bioRxiv 418376 (2018) https://doi.org/10.1101/418376
26. Data Type Registries Recommendations (Endorsed). Research Data Alliance. https://www.rd-alliance.org/group/data-type-registries-wg/outcomes/data-type-registries
27. GO FAIR International Support and Coordination Office (GFISCO). http://go-fair.org
28. Kahn, R., Wilensky, R.: A framework for distributed digital object services (1995). http://www.cnri.reston.va.us/k-w.html
29. Kahn, R., Wilensky, R.: A framework for distributed digital object services. Int. J. Digit. Libr. 6(2), 115–123 (2006). https://doi.org/10.1007/s00799-005-0128-x. https://www.doi.org/topics/2006_05_02_Kahn_Framework.pdf
30. Wittenburg, P., Strawn, G., Mons, B. et al.: Digital objects as drivers towards convergence in data infrastructures (2019). http://doi.org/10.23728/b2share.b605d85809ca45679b110719b6c6cb11
31. Object Oriented Programming. https://de.wikipedia.org/wiki/Objektorientierte_Programmierung
32. Liskov, B., Zilles, S.N.: Programming with abstract data types. In: ACM SIGPLAN Notices, vol. 9, no. 4, pp. 50–59. ACM, New York (1974)
33. Objet Storage. https://en.wikipedia.org/wiki/Object_storage
34. Handle System. https://en.wikipedia.org/wiki/Handle_System
35. GEDE Digital Object Topic Group. https://rd-alliance.org/group/gede-group-european-data-experts-rda/wiki/gede-digital-object-topic-group
36. GEDE Workshop on Digital Objects. https://rd-alliance.org/group/gede-group-european-data-experts-rda/wiki/first-gede-do-workshop-september-18

37. American Geophysical Union's Enabling FAIR Data Project. http://www.copdess.org/enabling-fair-data-project/
38. Supporting FAIR Exchange of Chemical Data Through Standards Development. GO FAIR Chemistry Implementation Network (ChIN). https://iupac.org/event/supporting-fair-exchange-chemical-data-standards-development/
39. Wilkinson, M.D., et al.: Interoperability and FAIRness through a novel combination of Web technologies. PeerJ Comput. Sci. **3**, e110 (2017). https://doi.org/10.7717/peerj-cs.110
40. FAIR Data Point Specification. https://github.com/DTL-FAIRData/FAIRDataPoint/wiki/FAIR-Data-Point-Specification
41. 500,000 data scientists needed in European open research data. JoinUp Platform, European Commission (2016). https://joinup.ec.europa.eu/news/500000-data-scientists-need
42. The Carpentries. https://carpentries.org
43. GO FAIR Implementation Networks. https://www.go-fair.org/implementation-networks/
44. GO FAIR Current Implementation Networks. https://www.go-fair.org/implementation-networks/overview/

Extensible Unifying Data Model Design for Data Integration in FAIR Data Infrastructures

Sergey Stupnikov$^{(\boxtimes)}$ and Leonid Kalinichenko

Institute of Informatics Problems, Federal Research Center
"Computer Science and Control" of the Russian Academy of Sciences,
Vavilova st. 44-2, 119333 Moscow, Russia
sstupnikov@ipiran.ru

Abstract. According to the Open Science paradigm data sources are to be concentrated within research data infrastructures intended to support the whole cycle of data management and processing. FAIR data management and stewardship principles that had being developed and announced recently state that data within a data infrastructure have to be findable, accessible, interoperable and reusable. Note that data sources can be quite heterogeneous and represented using very different data models. Variety of data models includes traditional relational model and its object-relational extensions, array and graph-based models, semantic models like RDF and OWL, models for semi-structured data like NoSQL, XML, JSON and so on. This particular paper overviews data model unification techniques considered as a formal basis for (meta)data interoperability, integration and reuse within FAIR data infrastructures. These techniques are intended to deal with heterogeneity of data models and their data manipulation languages used to represent data and provide access to data in data sources. General principles of data model unification, languages and formal methods required, stages of data model unification are considered and illustrated by examples. Application of the techniques for data integration within FAIR data infrastructures is discussed.

Keywords: FAIR data infrastructures · Data model unification ·
Data integration

1 Introduction

Data sources nowadays are quite heterogeneous: they are represented using very different data models. Variety of data models includes traditional relational model and its object-relational extensions, array and graph-based models, semantic models like RDF and OWL, models for semi-structured data like NoSQL, XML, JSON and so on. These models provide also very different data manipulation and query languages for accessing and modifying data.

According to Open Science paradigm [27] data sources are to be concentrated within research data infrastructures allowing access to data, computing services and processes. These infrastructures are intended to support the whole cycle of data management and

© Springer Nature Switzerland AG 2019
Y. Manolopoulos and S. Stupnikov (Eds.): DAMDID/RCDL 2018, CCIS 1003, pp. 17–36, 2019.
https://doi.org/10.1007/978-3-030-23584-0_2

processing from harvesting and curation to storage and analysis. Examples of research infrastructures are GEANT[1] (interconnecting Europe's national research and education networking organizations with high speed and bandwidth), EGI[2] (cloud and grid computing services), PRACE[3] (high-performance computing), IDGF[4] (desktop grid federation for crowd computing), OpenAIRE[5] (publications and research data storage), EUDAT[6] (collaborative data infrastructure for synchronization, exchange, store, share, search, replicate and get research data to computation), ELIXIR[7] (a distributed infrastructure for life-science information), EOSC[8] (European Open Science Cloud - a cloud for research data in Europe).

However, existing research data infrastructures rarely provide means for extracting maximum benefit from research investments. To force data and infrastructure providers to overcome this deficiency, for last several years data management and stewardship principles had being developed and announced [1]. These principles are called *FAIR Data Principles*. According to FAIR principles, data have to be findable, accessible, interoperable and reusable. Outcomes from data management and stewardship in the FAIR way are facilitating and simplifying the process of discovery, evaluation, and reuse of data within research infrastructures.

This paper overviews *data model unification* techniques considered as a formal basis for (meta)data interoperability, integration and reuse within FAIR data infrastructures. These techniques are intended to deal with heterogeneity of data models and their data manipulation languages used to represent data and provide access to data in data sources.

The main ideas of data model unification are as follows. The kernel of *unifying data model* (called *canonical*) has to be chosen for a data infrastructure. The canonical data model serves as the language for knowledge representation mentioned in FAIR I1 principle ((meta)data use a formal, accessible, shared, and broadly applicable language for knowledge representation) [1, 2]. Every *source data model* used to represent some set of sources within the infrastructure has to be mapped into the canonical model. Mapping can be accompanied with the extension of the canonical model kernel if required. A mapping should be formalized and verified: a formal proof that the mapping preserves semantics of data structures and data manipulation operations of the source data model should be provided.

As the kernel of the canonical model some concrete data model like SQL (conforming to ISO/ANSI SQL standard of 2011 or later) or RDF/RDF Schema with SPARQL query language can be used. To cover features of various source data models the canonical model has to be extensible. Examples of extensions are specific data

---

[1] https://www.geant.org/.

[2] https://www.egi.eu/.

[3] http://www.prace-ri.eu/.

[4] http://desktopgridfederation.org/.

[5] https://www.openaire.eu/.

[6] https://eudat.eu/.

[7] https://www.elixir-europe.org/.

[8] https://ec.europa.eu/research/openscience/index.cfm?pg=open-science-cloud.

structures (data types), compound operations or restrictions (dependencies). An extension is constructed for every source data model. Canonical model is formed as the union of the kernel data model and all extensions.

Data model unification techniques were extensively studied at FRC CSC RAS [3]. As the kernel of the canonical model specific object-frame language with broad range of modeling facilities was used [4]. Approaches for mapping of different classes of source data models were developed: process models [5], semantic models [6, 13], array [9] and graph-based [10] models, some other kinds of NoSQL models [8]. Techniques for verification of mappings applying a formal language based on the first order logic and set theory and supported by automatic and interactive provers were developed [11, 12].

Source data models unification and construction of a canonical data model is a prerequisite for data integration and reuse within a data infrastructure that may combine virtual data integration facilities (subject mediators) as well as data warehouses to integrate heterogeneous data sources in an interoperable way [25].

The rest part of the paper is structured as follows: Sect. 2 overviews data unification techniques that have been developed during recent years and Sect. 3 discusses application of these techniques for data integration within FAIR data infrastructures.

This work is an extension of [26]. Comparing it with the previous work in [26], introduction was extended, Sects. 2 and 3 were significantly extended with details and examples of data unification techniques.

2 Data Model Unification

2.1 General Principles of Data Model Unification

Various source data models and their data manipulation languages applied within some data infrastructure have to be unified in the frame of some canonical data model.

The main principle of the canonical model design (synthesis) for a data infrastructure is the *extensibility* of the canonical model kernel in heterogeneous environment [3], including various models used for the representation of sources of the data infrastructure. A kernel of the canonical model is fixed. A specific source data model R of the environment is said to be *unified* if it is mapped into the canonical model C [11, 12]. This means a creation of such extension E of the canonical model kernel (note that such extension can be empty) and such mapping M of a source model into extended canonical one that the source model *refines* the extended canonical one. Model refinement of C by R means that for any admissible specification (schema) r represented in R its image $M(r)$ in C under the mapping M is refined by the specification r. Such refining mapping of models means preserving of operations and information of a source model after mapping it into the canonical one. Preserving of operations and information should be formally proven. The canonical model for the environment is synthesized as the union of extensions, constructed for all models of the environment (Fig. 1).

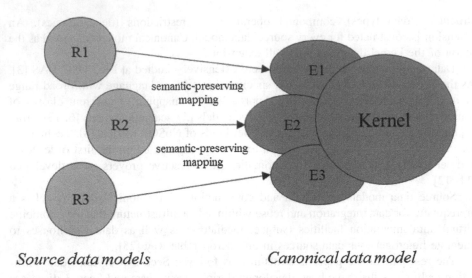

Source data models   *Canonical data model*

Fig. 1. Extensible unifying data model design

2.2   Languages and Formal Methods Required for Data Model Unification

The following languages and formal methods are required to support data model unification:

- a kernel of the canonical data model;
- formal methods allowing to describe data model syntax as well as semantic mappings (transformations) of one model to another;
- formal methods supporting verification of refinement reached by the mapping.

Within studies on data unification techniques at FRC CSC RAS as *a kernel of the canonical data model* the SYNTHESIS language [4] was used. The SYNTHESIS language, as a hybrid semistructured and object-oriented data model, includes the following distinguishing features: facilities for definitions of frames, abstract data types, classes and metaclasses, functions and processes, logical formulae facilities applied for description of constraints, queries, pre- and post-conditions of functions, assertions related to processes. For extension of the canonical model kernel, metaclasses, meta-frames, parameterized constructions including assertions and generic data types were applied. Data unification teqhniques developed can be adopted also for other canonical data model kernels like SQL or RDF.

For *data model's semantics formalization and refinement verification* the AMN (Abstract Machine Notation) language [14] was used. The language is supported by technology and tools for proving of refinement (B-technology) [15]. AMN is based on the first order predicate logic and Zermelo-Frenkel set theory and enables to consider state space specifications and behavior specifications in an integrated way. The system state is specified by means of state variables and invariants over these variables, system behavior is specified by means of operations defined as generalized substitutions – a sort of predicate transformers. Refinement of AMN specifications is formalized as a set of refinement proof obligations – theorems of first order logic. Generally speaking in terms of pre- and post-conditions of operations, refinement of AMN specifications means weakening pre-conditions and strengthening post-conditions of corresponding operations included in these specifications. Proof obligations are generated automatically and should be proven with the help automatic and interactive theorem prover, for instance, Atelier [15].

For the *formal description of model syntax and transformations* two approaches were developed and prototyped.

The first approach [11, 12] is based on the metacompilation languages SDF (Syntax Definition Formalism) and ASF (Algebraic Specification Formalism). For the languages a tool support—Meta-Environment [16]—is provided based on term rewriting techniques. Data model syntax is represented using SDF in a version of extended Backus–Naur form. Data model transformations are defined as ASF language modules which are sets of functions. A function defines a transformation of a syntactic element of a source model into a syntactic element of the canonical model. Recursive calls of transformation functions are allowed. According to the ASF-definition the transformation program code (C language) is generated automatically by means of Meta-Environment tools. The transformation obtained is used for mapping of source model specifications into the canonical model specifications.

The second approach [17] is based on the Model-Driven Architecture (MDA) [18] proposed by Object Management Group. Data model abstract syntax neglecting any syntactic sugar is defined using *Ecore* metamodel (an implementation of OMG's Essential Meta-Object Facility) used in Eclipse Modeling Framework [19]. Concrete syntax of data models binding syntactic sugar and abstract syntax is formalized using EMFText framework [20]. Data model transformations are defined using ATLAS Transformation Language (ATL) [21] combining declarative and imperative features. ATL transformation programs are composed of rules that define how source model elements are matched and navigated to create and initialize the elements of the target models. Type system of the ATL is very close to the type system of the OMG Object Constraint Language.

2.3   Stages of Data Model Unification

Construction of a mapping of a source data model R into the canonical model C is divided into the following stages:

- definition of reference schemas of the models R and C (if the latter has not yet been defined);
- integration of reference schemas of the model R and C;
- syntax formalization for the models R and C (if the latter has not yet been defined);
- creation of a required extension E of the canonical model C;
- construction of a transformation of the model R into the extended canonical model;
- formalization of the data models semantics and verification of refinement of the extended canonical model by the model R.

All stages of data model unification are illustrated below with examples.

Definition of Reference Schemas. *The Reference schema of a data model* is an abstract description containing concepts related to constructs of the model and significant associations among these concepts. Using MDA terms reference schemas are just metamodels conforming the Ecore metamodel [19]. As an example, Fig. 2 shows several elements of the reference schema of the OWL language considered as a source data model (left hand part of the figure) and several elements of reference schema of the SYNTHESIS language considered as the canonical data model (right hand part of the figure). The OWL reference schema is developed on the basis of W3C Recommendation [22] and the SYNTHESIS reference schema is developed on the basis of description of the language [4]. OWL concepts shown are *RDFSClass* and its subconcept *OWLClass* accompanied by *uriRef* attribute and *subClassOf* and *disjointClass* associations. SYNTHESIS concepts relevant to *OWLClass* are *ADTDef* (abstract data type definition) and *ClassDef* (class definition) accompanied by *name* attribute and *supertypes* and *superclasses* associations.

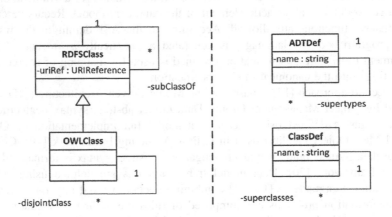

Fig. 2. Reference schema elements for OWL and SYNTHESIS

Table 1. Correspondences between the OWL and the SYNTHESIS elements

OWL element	SYNTHESIS element
OWLClass	ADTDef, ClassDef
OWLClass.uriRef	ADTDef.name, ClassDef.name
OWLClass.subclassOf	ADTDef.supertypes, ClassDef.superclasses

Integration of Reference Schemas. The aim of the integration of the reference schemas is to identify the relevant constructions of source and canonical models. The list of correspondences between OWL and the SYNTHESIS elements for the subsets of the reference schemas shown on the Fig. 2 is presented in the Table 1. The integration can be partially automatized if reference schema elements are provided with verbal definitions. However, in most cases correspondences have to be established and confirmed by the expert.

Data Model Syntax Formalization. Syntax formalization and binding the syntax with reference schema using EMFText framework for the SYNTHESIS model elements shown on Fig. 2 are illustrated below:

```
SYNTAXDEF syn FOR http://synthesis.ipi.ac.ru/Synthesis/
RULES
ADTDef ::=
  "{" name[]  ";"  "in" ":" "type"  ";"
  ("supertypes" ":"  supertypes[] (","  supertypes[])* ";")?    "}" ;
ClassDef ::=
  "{" name[]  ";"  "in" ":" "class"  ";"
  ("superclass" ":"  superclasses[] (","  superclasses[])* ";")?   "}" ;
```

For a concept of the reference schema (like *ADTDef* and *ClassDef*) a grammar rule is defined. The rule determines general syntactic structure of the element using

- terminal symbols embraced by quotes (like *"supertypes"* or *"{"*);
- nonterminal symbols corresponding to attributes and associations of the concept (like *name* or *supertypes*);
- grammar constructions denoting multiplicity (*), optionality (?), and choice (|).

Syntax formalization with SDF looks almost the same as with EMFText except minor syntactic differences.

Creation of an Extension of the Canonical Model. An example of extension of the SYNTHESIS language as a canonical model for unification of the OWL language is *Transitive* association metaclass:

```
{ Transitive; in: association, metaclass;
instance_section: {
  domain: type; range: type;
  transitivity: { in: predicate, invariant;
  {{ all a,b,c(in([a,b], this) & in([b,c], this) -> in([a,c], this)) }}
} } }
```

The extension is required to reason about *transitive object properties* defined in OWL [22]. Association metaclass [4] is a collection of associations. An association is a set of associated pairs of objects. Association transitivity is expressed by the invariant *transitivity* of the metaclass stating the following. Let *a*, *b*, *c* be objects. If *a* is associated with *b* according to association *assoc* and *b* is associated with *c* according to *assoc* then *a* is associated with *c* according to *assoc*. So any association defined to be an instance of *Transitive* metaclass is transitive. In metaclass specification above *this* refers to current instance, *in* denotes set membership predicate, *[a, b]* denotes a pair of associated objects. Invariant specification is defined by a first order logic formula: *all* denotes universal quantifier, & denotes conjunction, -> denotes implication. More details concerning this extension can be found in [12].

Another example of extension of the SYNTHESIS language as a canonical model for unification of an attributed graph data model is considered in [10]. In particular, the extension includes *vertices* and *edges* classes intended to represent vertices of a graph and edges of a graph respectively. Specification of *edges* class looks as follows (only a part of specification is considered):

```
{ edges; in: class;
  instance_section: {
    startVertex: vertices.inst;
    endVertex: vertices.inst;
    isValidEdge: { in: predicate;
      params: {+stVtx/vertices.inst, +endVtx/vertices.inst,
               returns/Boolean };
      {{ (stVtx = this.startVertex &
           endVtx = this.endVertex -> returns = true) &
         (stVtx <> this.startVertex | endVtx <> this.endVertex) ->
           returns = false) }}
    };
  }
```

Any edge of a graph is an instance of the *edge* class. Every edge has its head (*startVertex*) and tail (*endVertex*) vertices. A predicate *isValidEdge* is defined over pairs of vertices. For an edge *e* the predicate *e.isValidEdge(v1, v2)* turns true if and only if the head of *e* (*e.startVertex*) equals to *v1* and the tail of *e* (*e.endVertex*) equals to *v2*.

Construction of a Transformation of a Source Model into the Extended Canonical Model. As mentioned in Sect. 2.2 model transformations are proposed to be constructed in two ways. The first way applies the metacompilation language ASF (Algebraic Specification Formalism). The second way applies ATLAS Transformation Language (ATL). Both ways are illustrated below.

ASF transformation constructed to transform OWL model elements into the SYNTHESIS model elements shown on Fig. 2 looks as follows:

```
module unifier/owl2synthesis/owl-translator
imports unifier/owl/OWL-Syntax
imports unifier/synthesis/Synthesis-Syntax
context-free syntax
t-Type-Specification-List(Directive*, Directive*) ->
   {Type-Specification ","}*
t-Class-Declarator-List(Directive*) -> {Class-Declarator ","}*
variables
"Directive*"[0-9\']* -> Directive*
"Type-Specification*"[0-9\']* -> {Type-Specification ","}*
equations
Type-Specification* :=
   t-Type-Specification-List(Directive*, Directive*2),
Synthesis-Id := t-Synthesis-Id(OwlID)
====>
t-Type-Specification-List(
   Class(OwlID Description*)
   Directive*,
   Directive*2
) =
{ Synthesis-Id; in: type, owl;
   t-Type-Supertype-Section(Description*)
},
Type-Specification*


Synthesis-Id := t-Synthesis-Id(OwlID)
====>
t-Class-Declarator-List(
   Class(OwlID Description*)
   Directive*
) =
{ Synthesis-Id; in: class, owl;
   instance_section: Synthesis-Id;
},
t-Class-Declarator-List(Directive*)
```

The transformation imports syntax definitions for source (*OWL-Syntax*) and canonical model (*Synthesis-Syntax*). Two transformation functions are defined: the first is required to transform OWL class specifications into the SYNTHESIS type specifications (*t-Type-Specification-List*), and the second is required to transform OWL class specifications into the SYNTHESIS class specifications (*t-Class-Declarator-List*). Function signatures are defined in *context-free syntax* section of the transformation. Transformation rules are defined as term rewritings in *equations* section. Both rules include matching conditions of the form $A := B$ and term equations. Both matching conditions and equalities contain recursive applications of transformation functions. Additional syntactic variables used in transformation rules are declared in *variables* section.

ATL transformation constructed to transform OWL model elements into the SYNTHESIS model elements shown on Fig. 2 looks as follows:

```
module OWL2Synthesis;
create OUT : Synthesis from IN : OWL;
rule OWLClass{
  from c: OWL!OWLClass
  using{
    sup: Set(OWL!OWLClass) =
      c.subClassOf->select(e | e.oclIsTypeOf(OWL!OWLClass)); }
  to
    type: Synthesis!ADTDef(
      name <- c.resourceName(),
      supertypes <- sup),
    class: Synthesis!ClassDef(
      name <- c.resourceName().toLower(),
      instanceType <- type)
  do{
    class.superclasses <-
      sup->collect(e | thisModule.resolveTemp(e, 'class'));
  }
}
```

For a source model element type (*OWLClass*) a transformation rule with the same name is defined. Source element of the rule (*c*) is declared and typed in *from* section of the rule. Additional variables are declared and initialized in *using* section. For instance, *sup* variable is initialized as a set of superclasses of the source class *c*. Target elements (*type, class*) are declared and typed in *to* section. For a target element a set of bindings is specified. Bindings define the way the features (either attributes or references like *name, supertypes, instanceType, superclasses*) of the generated element must be initialized. Bindings are defined either in declarative way in *to* section or in imperative way in *do* section.

For both ASF and ATL transformations their *templates* can be generated automatically on the basis of correspondences between a source and the canonical model elements. Some details concerning ASF template generation can be found in [11, 12]. Details concerning ATL template generation can be found in [17]. Anyway, transformations have to be completed (extended, modified) by an expert.

Formalization of Data Model Semantics and Verification of Data Model Refinement can be performed in two ways.

In the first way formalization of data model semantics means a construction of transformations of source and canonical data model specifications into AMN-specifications. These semantic transformations are constructed manually by an expert applying ASF or ATL languages similarly to transformations considered in the previous subsubsection. Additional technical details are omitted here, only results of semantic transformations application are illustrated by examples.

Applying semantic transformation for any specification of a source data model the AMN-specification expressing its semantics is generated automatically. For instance, a tiny OWL specification and its semantic representation are shown in Table 2.

Source OWL specification includes *Person* class and a transitive property *hasAncestor*. A set of individuals of the ontology is represented in AMN by the set *Ind*. *Person* class is represented by a variable typed in invariant as subset of *Ind*. Object property *hasAncestor* is represented by a variable typed as total function with domain and range corresponding to domain and range of the property.

Table 2. Source specification and its semantics in AMN

OWL Specification	AMN Specification

`Ontology(`
`<http://example.com/`
`owl/families>`
`Declaration(`
`  Class(Person))`
`Declaration(`
`  ObjectProperty(`
`    hasAncestor))`
`TransitiveObjectProp-`
`erty( hasAncestor )`
`ObjectPropertyDomain(`
`  hasAncestor Person)`
`ObjectPropertyRange(`
`  hasAncestor Person))`
	`REFINEMENT Families`
	`SETS Ind`
	`ABSTRACT_VARIABLES Person, hasAncestor`
	`INITIALISATION`
	`Person := {} ||`
	`hasAncestor := {}`
	`INVARIANT Person: POW(Ind) &`
	`hasAncestor: Person --> POW(Person) &`
	`!(aa, bb, cc).(aa: Ind & bb: Ind & cc: Ind &`
	`  bb: hasAncestor(aa) & cc: hasAncestor(bb) =>`
	`  cc: hasAncestor(aa))`
	`OPERATIONS`
	`add_hasAncestor(ind, val) =`
	`PRE ind: ext_Person & val: ext_Person`
	`THEN`
	`    hasAncestor(ind) := hasAncestor(ind) \/`
	`    {val} \/ hasAncestor(val)`
	`END`
	`END`

Transitivity of the property is represented by a conjunctive part of the invariant. The property is represented also by the operation *add_hasAncestor* adding an ancestor *val* of an individual *ind*. To preserve transitivity every ancestor of *val* becomes also ancestor of *ind*.

Applying *OWL2Synthesis* transformation illustrated above *families* ontology can be transformed into the SYNTHESIS module specification with the same name (Table 3, left hand part) including abstract type *Person* with *hasAncestor* transitive association attribute and class *person* with instance type *Person*. Using semantic transformation of the SYNTHESIS language [23] the respective AMN-specification for the module is generated automatically (Table 3, right hand part).

General principles for representation structures of the SYNTHESIS language in AMN are overviewed further [23]. Set of all abstract values (values of all abstract data types) is represented in AMN by the *AVAL* set. Set of object type values is represented by the set *Obj* that is a subset of *AVAL*. Type *Person* is represented by its extent *ext_Person*, that is the set of admissible values. Extent is typed in PROPERTIES clause as subset of *Obj*. Class *person* is represented by a variable with the same name typed in INVARIANT clause as subset of extent of its instance type *Person*. Attribute association *hasAncestor* is represented by a variable with the same name typed in INVARIANT clause as a function having type extent as the domain. Association metaclass *Transitive* is also represented by a variable and typed in invariant alongside with transitive property represented by the respective first order logic formula. Attribute association *hasAncestor* is defined to be an instance of *Transitive* set. The attribute association is represented also by the operation *add_hasAncestor* adding an ancestor *val* of a person *ind*.

Table 3. Canonical specification and its semantics in AMN

SYNTHESIS Specification	AMN Specification

{ families;
  in: module,
    ontology;
 	
type:
{ Person;
  in: type;
  hasAncestor:
    Person;
  metaslot
    in: Transitive;
  end
};
 	
class_specification:
{ person;
  in: class;
  instanceType:
    Person;
};
}
	REFINEMENT CanonicalFamilies
	SETS AVAL
	ABSTRACT_CONSTANTS Obj, ext_Person
	PROPERTIES
	Obj: POW(AVAL) & ext_Person: POW(Obj)
	ABSTRACT_VARIABLES
	Transitive, person, hasAncestorCan
	INITIALISATION
	Transitive := {} ||
	person := {} || hasAncestorCan := {}
	INVARIANT
	Transitive: POW(Obj +-> POW(Obj)) &
	!(tt, aa, bb, cc).(tt: Transitive & aa: Obj &
	    bb: Obj & cc: Obj & bb: tt(aa) & cc: tt(bb) =>
	    cc: tt(aa)) &
	person: POW(ext_Person) &
	hasAncestorCan: ext_Person --> POW(ext_Person) &
	hasAncestorCan: Transitive
	OPERATIONS
	add_hasAncestor(ind, val) =
	PRE ind: ext_Person & val: ext_Person
	THEN
	  hasAncestorCan(ind) := hasAncestorCan(ind) \/
	{val} \/ hasAncestorCan(val)
	END
	END

Having AMN semantic specifications for both source and canonical specifications, refinement of the canonical data model specification *families* by a source ontology *families* is reduced to refinement of their semantic AMN specifications and can be verified applying the refinement theorem prover Atelier [15]. For this particular example, *linking invariant* [14] which is aimed to bind variables of refined specification *CanonicalFamilies* (*person, hasAncestorCan*) with variables of refining specification *Families* (*Person, hasAncestor*) was added into *Families* specification:

```
Person = person &
!(ind, val).(ind: Ind & val: POW(Ind) =>
          ((hasAncestor(ind) = val) <=> (hasAncestorCan(ind) = val)))
```

Specification refinement w.r.t. linking invariant was automatically reduced to 20 theorems, 9 of them were proved automatically. Generally speaking, verification of data model refinement is realized over a set of source model specification samples.

In the second way semantics of a data model (source or canonical) as a whole is expressed by an AMN specification. For instance, in [9] AMN semantics for an array data model is defined, in [10] AMN semantics for a graph data model is defined. AMN semantics for the SYNTHESIS language as the canonical data model was also provided [9, 10].

A subset of AMN semantic specification for a graph data model looks as follows:

```
REFINEMENT GraphDM
ABSTRACT_VARIABLES attributeIDs, attributes, attributeTyping, vertices,
    edges, headVertix, tailVertix, g_integerAttributeValue
INVARIANT
vertexTypeIDs: POW(NAT) & edgeTypeIDs: POW(NAT) &
attributeIDs: POW(NAT) &
attributes: vertexTypeIDs \/ edgeTypeIDs --> POW(attributeIDs) &
attributeTyping: attributeIDs --> BuiltInTypes &
vertices: POW(NAT) & edges: POW(NAT) &
headVertix: edges --> vertices & tailVertix: edges --> vertices &
g_integerAttributeValue: (vertices \/ edges)*attributeIDs +-> INT
OPERATIONS
deleteVertex(attr, cond) =
PRE attr: attributeIDs & cond: INT --> BOOL &
    attributeTyping(attr) = Integer
THEN
vertices := vertices -
{vert | vert: vertices & attr: attributes(vertixType(vert)) &
    cond(g_integerAttributeValue(vert, attr)) = TRUE }
END
END
```

Vertices of graphs are represented by the *vertices* variable, edges are represented by the *edges* variable. Bindings of edges and vertices are represented by *headVertix* and *tailVertix* variables. Set of attribute identifiers is represented by the *attributeIDs* variable, types of attributes are represented by the *attributeTyping* variable. Bindings of attributes with vertices and edges are represented by the *attributes* variable. Integer attribute values are represented by the *g_integerAttributeValue* variable (attribute values of other types are represented by similar variables). A data manipulation operation *deleteVertex(attr, cond)* is defined. The operation removes all vertices *vert* from the set *vertices* such that integer attribute *attr* is defined for *vert* and its value satisfies the condition *cond*.

A subset of the AMN semantic specification for the SYNTHESIS language extended for unification the graph data model looks as follows:

```
REFINEMENT ObjectDM
CONSTANTS
c_edges, c_vertices, a_startVertex, a_endVertex
PROPERTIES
c_edges: STRING_Type & c_vertices: STRING_Type &
a_startVertex: NAT & a_endVertex: NAT
ABSTRACT_VARIABLES
classNames, attributeNames, attributeType,
objectIDs, objectsOfClass,
adtAttributeValue, integerAttributeValue, isValidEdge
INVARIANT
classNames: POW(STRING_Type) &
attributeNames: NAT +-> STRING_Type &
attributeType: dom(attributeNames) +-> BuiltInTypes &
objectIDs: POW(NAT) & objectsOfClass: classNames --> POW(objectIDs) &
integerAttributeValue: dom(attributeNames) +-> (objectIDs +-> INT) &
adtAttributeValue: dom(attributeNames) +-> (objectIDs +-> NAT) &
isValidEdge:
   objectsOfClass(c_vertices)*objectsOfClass(c_vertices) --> BOOL &
!(edg, v1, v2).(edg: objectsOfClass(c_edges) &
  v1: objectsOfClass(c_vertices) & v2: objectsOfClass(c_vertices) =>
  ((isValidEdge(v1, v2) = TRUE) <=>
    (adtAttributeValue(a_startVertex)(edg) = v1 &
     adtAttributeValue(a_endVertex)(edg) = v2) ) )
OPERATIONS
deleteVertex(attr, cond) =
PRE   attr : dom(attributeNames) & cond : INT --> BOOL &
      attributeType(attr) = Integer
THEN
objectsOfClass(c_vertices) :=
objectsOfClass(c_vertices) -
{ vert | vert: objectsOfClass(c_vertices) &
       vert: dom(adtAttributeValue(attr)) &
       cond(integerAttributeValue(attr)(vert)) = TRUE }
END
END
```

Constants required for unification of the graph data model are declared in CONSTANTS clause and typed in PROPERTIES clause. So *c_edges* is the name of the class for edges, *c_vertices* is the name of the class for vertices, *a_startVertex* is the identifier of the attribute representing heads of edges, and *a_endVertex* is the identifier of the attribute representing tails of edges. Sets of class and attribute names are represented by the *classNames* and *attributeNames* variables respectively. Set of object identifiers is represented by the *objectIDs* variable. Bindings of classes with their objects are represented by the *objectsOfClass* variable. Attributes are bound with their types via the *attributeType* variable. Attribute names are bound with objects and attribute values via *adtAttributeValue* and *integerAttributeValue* variables. Variable *isValidEdge* represents the respective predicate defined in instance type of the *edge* class as an extension of the SYNTHESIS language considered earlier. Operation *deleteVertex* is defined to unify the respective operation of the graph data model.

As far as semantic AMN specifications for the graph data model and extended canonical data model are defined, refinement of the canonical model by the graph data model required for unification is reduced to the refinement of their semantic specifications. Linking invariant that binds variables of refining and refined specifications was constructed and proof of refinement was performed using Atelier B, additional details can be found in [10]. An example of a predicate constituting the linking invariant looks as follows:

```
!edg.(edg: edges =>
   headVertix(edg) = adtAttributeValue(a_startVertex)(edg) &
   tailVertix(edg) = adtAttributeValue(a_endVertex)(edg)   )
```

The predicate binds *headVertix* and *tailVertix* variables of the *GraphDM* specification with *a_startVertex* and *a_endVertex* variables of the *ObjectDM* specification.

Partial Automation of Data Model Unification Techniques mentioned above was implemented within Unifying Information Models Constructor (Model Unifier in short) [11, 12]. Unifier consists of the following main components:

- tool for the formal description and correctness checking of model syntax and transformations (Meta-Environment, ATL Tools);
- Atelier [15], supporting AMN and providing facilities for proving of specification refinement;
- model manager.

Meta-Environment, ATL Tools and Atelier B are third-party products. Model manager provides a graphical interface allowing an expert to search for, view and register data models and extensions of the canonical model; to call specific components for generating templates, editing and integration of reference schemas, generating templates for translators of source models into the canonical one, translation of source models specifications into AMN or into canonical specifications, translation of canonical specifications into AMN.

Application of Data Model Unification Techniques. Recent years data model unification techniques were applied to wide range of source data models. In [5] a canonical process model has been synthesized for the environment of workflow patterns classified by W. M. P. van der Aalst. Thus the canonical process model possesses a property of completeness with respect to broad class of process models used in various Workflow Management Systems as well as the languages used for process composition of Web services.

In [11, 12] the Ontology Web Language was unified with the SYNTHESIS language, in [6] OWL 2 QL was mapped into the SYNTHESIS.

In [7] application of the canonical model synthesis methods for the *value inventive data models* was discussed. The distinguishing feature of these data models is inference of new, unknown values in the process of query answering.

In [8] an approach to mapping of different types of NoSQL models into the object model of the SYNTHESIS language used as unifying data model was considered.

In [9] unification of an array-based data model used in SciDB DBMS was considered, and in [10] unification of an attributed graph data model was considered. For both models verification using AMN specifications is provided.

In [13] issues on unification of RDF with accompanying RDF Schema and SPARQL languages were discussed.

3   FAIR Data Based on Data Model Unification

The following levels of integration (from higher to lower) can be distinguished: data model integration (unification), schema matching and integration (metadata integration) and data integration proper. On the level of data models elements of models (languages) are matched. On the level of schemas data types (structures) and their attributes are matched. On the level of data proper rules for transformation of data collections, their elements and attribute values are defined. An example of model element, schema element, and data attribute value matching is shown on Fig. 3. On the data model level the UML class is matched with the table of the relational data model and its attributes are matched with columns of the table. On the schema level *Books* type of source schema is matched with *BookInfo* type of target schema, attributes are matched respectively. On the level of data proper two source complementary descriptions of digital camera are fused to form the target description.

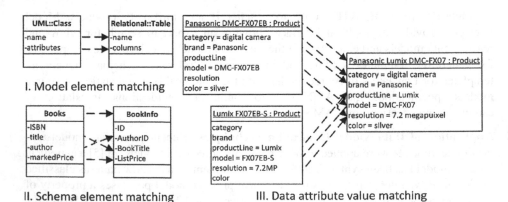

I. Model element matching

II. Schema element matching   III. Data attribute value matching

Fig. 3. Model, schema, and data element matching for data integration

Usually completion of the integration on a higher level is a prerequisite for integration on a lower level. Obviously the highest level, i.e. data model unification is a prerequisite for (meta)data interoperability, integration and reuse within FAIR data infrastructures and data model unification techniques overviewed in the previous section can be considered as a formal basis for achieving FAIRness of data.

Any level of integration makes data more FAIR: integrated data are much easier to find, access and reuse and also integrated data are more interoperable than heterogeneous data stored in different data sources. The most mature level of integration is achieved within data integration systems like *subject mediators* or *data warehouses*.

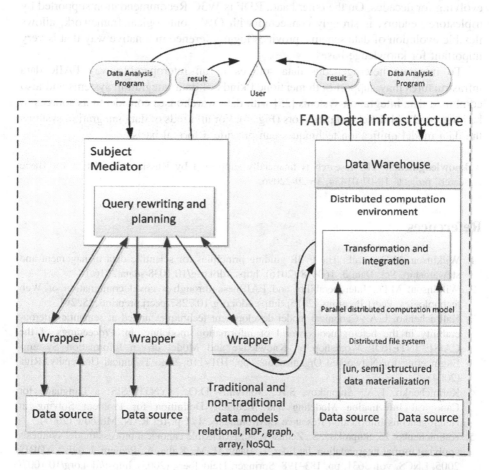

Fig. 4. FAIR data infrastructure combining virtual and materialized data integration

Subject mediators implement *virtual integration* with user queries defined in some unified data model. Such queries are to be decomposed into sets of subqueries and these subqueries are to be transferred to heterogeneous data sources. Data sources are connected with a subject mediator via wrappers which transforms queries into source data models and also transforms query answers from source data models into unified mediator data model. Query answers are transferred by wrappers back to the mediator, combined and sent to users. One of the latest trends nowadays is construction of subject mediators over *data lakes* [24].

Data warehouses implement *materialized integration* with all required data extracted from sources, transformed into unified warehouse data model, and stored into a warehouse.

Any kind of integration system requires unified data model. One of the important issues to be resolved for data integration within FAIR data infrastructures is the choice of the canonical model kernel. Even the choice between SQL and RDF is difficult. On the one hand, SQL is supported by industrial standards, methods and technologies evolving for decades. On the other hand, RDF is W3C Recommendation supported by triplestore vendors, is strongly connected with OWL ontological framework, allows flexible evolution of data schema, provides logic inference in a native way that is very important for knowledge bases.

To integrate heterogeneous data sources in an interoperable way FAIR data infrastructures may support both mentioned kinds of data integration systems and also combined data integration systems [25] with data warehouses considered as sources to be integrated within subject mediators (Fig. 4). For all kinds of data integration systems the data model unification techniques can provide a formal basis.

Acknowledgments. The research is financially supported by Russian Foundation for Basic Research, projects 18-07-01434, 18-29-22096.

References

1. Wilkinson, M.D., et al.: The FAIR guiding principles for scientific data management and stewardship. Sci. Data **3**, 160018 (2016). https://doi.org/10.1038/sdata.2016.18
2. Wilkinson, M.D.: Interoperability and FAIRness through a novel combination of Web technologies. PeerJ Preprints (2016). https://doi.org/10.7287/peerj.preprints.2522v1
3. Kalinichenko, L.A.: Canonical model development techniques aimed at semantic interoperability in the heterogeneous world of information modeling. In: Proceedings of the CAiSE INTEROP Workshop on Knowledge and Model driven Information Systems Engineering for Networked Organizations, pp. 101–116. Riga Technical University, Riga (2004)
4. Kalinichenko, L.A., Stupnikov, S.A., Martynov D.O.: SYNTHESIS: a Language for Canonical Information Modeling and Mediator Definition for Problem Solving in Heterogeneous Information Resource Environments, 171 p. IPI RAN, Moscow (2007)
5. Kalinichenko, L., Stupnikov, S., Zemtsov, N.: Extensible canonical process model synthesis applying formal interpretation. In: Eder, J., Haav, H.-M., Kalja, A., Penjam, J. (eds.) ADBIS 2005. LNCS, vol. 3631, pp. 183–198. Springer, Heidelberg (2005). https://doi.org/10.1007/11547686_14
6. Kalinichenko, L.A., Stupnikov, S.A.: OWL as yet another data model to be integrated. In: Proceedings of the 15th East-European Conference on Advances in Databases and Information Systems, pp. 178–189. Austrian Computer Society, Vienna (2011)
7. Kalinichenko, L., Stupnikov, S.: Synthesis of the canonical models for database integration preserving semantics of the value inventive data models. In: Morzy, T., Härder, T., Wrembel, R. (eds.) ADBIS 2012. LNCS, vol. 7503, pp. 223–239. Springer, Heidelberg (2012). https://doi.org/10.1007/978-3-642-33074-2_17

8. Skvortsov N.A.: Mapping of NoSQL data models to object specifications. In: Proceedings of the 14th Russian Conference on Digital Libraries RCDL 2012. CEUR Workshop Proceedings, vol. 934, pp. 53–62 (2012)
9. Stupnikov, S.A.: Unification of an array data model for the integration of heterogeneous information resources. In: Proceedings of the 14th Russian Conference on Digital Libraries RCDL 2012. CEUR Workshop Proceedings, vol. 934, pp. 42–52 (2012)
10. Stupnikov, S.A.: Mapping of a graph data model into an object-frame canonical information model for the development of heterogeneous information resources integration systems. In: Proceedings of the 15th Russian Conference on Digital Libraries RCDL 2013. CEUR Workshop Proceedings, vol. 1108, pp. 85–94 (2013)
11. Zakharov, V.N., Kalinichenko, L.A., Sokolov, I.A., Stupnikov, S.A.: Development of canonical information models for integrated information systems. Inform. Appl. 1(2), 15–38 (2007)
12. Kalinichenko, L.A., Stupnikov, S.A.: Constructing of mappings of heterogeneous information models into the canonical models of integrated information systems. In: Proceedings of the 12th East-European Conference on Advances in Databases and Information Systems, pp. 106–122. Tampere University of Technology, Pori (2008)
13. Skvortsov N.A.: Mapping of RDF data model into the canonical model of subject mediators. In: Proceedings of the 15th Russian Conference on Digital Libraries RCDL 2013. CEUR Workshop Proceedings, vol. 1108, pp. 95–101 (2013)
14. Abrial, J.-R.: The B-Book: Assigning Programs to Meanings. Cambridge University Press, Cambridge (1996)
15. Atelier, B.: The industrial tool to efficiently deploy the B Method. http://www.atelierb.eu/
16. van den Brand, M.G.J., et al.: The ASF+SDF meta-environment: a component-based language development environment. In: Wilhelm, R. (ed.) CC 2001. LNCS, vol. 2027, pp. 365–370. Springer, Heidelberg (2001). https://doi.org/10.1007/3-540-45306-7_26
17. Stupnikov, S.A., Kalinichenko, L.A.: Methods for semi-automatic construction of information models transformations. In: Proceedings of the 13th East-European Conference Advances in Databases and Information Systems, workshop Model – Driven Architecture: Foundations, Practices and Implications (MDA), pp. 432–440. Riga Technical University, Riga (2009)
18. Object Management Group Model Driven Architecture (MDA): MDA Guide rev. 2.0. OMG Document ormsc/2014–06-01 (2014)
19. Steinberg, D., Budinsky, F., Paternostro, M., Merks, E.: EMF: Eclipse Modeling Framework, 2nd edn. Addison-Wesley Professional, Boston (2008)
20. EMFText Concrete Syntax Mapper. https://github.com/DevBoost/EMFText
21. ATL - a model transformation technology. https://eclipse.org/atl/
22. OWL 2 Web Ontology Language Structural Specification and Functional-Style Syntax (Second Edition). W3C Recommendation. https://www.w3.org/TR/owl-syntax/ (2012)
23. Stupnikov, S.A.: A semantic transformation of the canonical information model into a formal specification langage for the refinement verification. In: Proceedings of the 12th Russian Conference on Digital Libraries RCDL 2010, pp. 383–391. Kazan Federal University, Kazan (2010)
24. Hai, R., Quix, C., Zhou, C.: Query rewriting for heterogeneous data lakes. In: Benczúr, A., Thalheim, B., Horváth, T. (eds.) ADBIS 2018. LNCS, vol. 11019, pp. 35–49. Springer, Cham (2018). https://doi.org/10.1007/978-3-319-98398-1_3
25. Stupnikov, S.A., Vovchenko, A.E.: Combined virtual and materialized environment for integration of large heterogeneous data collections. In: 16th Russian Conference on Digital Libraries RCDL 2014 Proceedings. CEUR Workshop Proceedings, vol. 1297, pp. 201–210 (2014)

26. Stupnikov, S.: FAIR data based on extensible unifying data model development. Selected Papers of the XX International Conference on Data Analytics and Management in Data Intensive Domains (DAMDID/RCDL 2018), CEUR Workshop Proceedings, vol. 2277, pp. 9–13 (2018)
27. Salmi, J.: Study on open science: impact, implications and policy options (2015). https://ec.europa.eu/research/innovation-union/pdf/expert-groups/rise/study_on_open_science-impact_implications_and_policy_options-salmi_072015.pdf

Meaningful Data Reuse in Research Communities

Nikolay A. Skvortsov(✉) ⓘ

Institute of Informatics Problems, Federal Research Center "Computer Science
and Control" of Russian Academy of Sciences, Moscow, Russia
nskv@mail.ru

Abstract. FAIR data principles declare data interoperability and reuse
according to machine and human readable shared specifications. Adherence to
this set of principles brings some implications for data infrastructures and
research communities. Meaningful data exchange and reuse by humans and
machines require formal specifications of research domains accompanying data
and allowing automatic reasoning. Development of formal conceptual specifications in research communities can be stimulated by a necessity to reach
semantic interoperability of data collections and components, and reuse of data
resources. Usage of formal domain specifications reduces data heterogeneity
costs. Formal reasoning allows meaningful search and verified reuse of data,
methods, and processes from collections. These means can make research
lifecycle in communities more efficient. A lifecycle includes collecting domain
knowledge specifications, classifying all data, methods, and processes according
to such specifications, reusing relevant data and methods, and collecting and
sharing results for reuse.

Keywords: FAIR data principles · Research data infrastructure ·
Research community · Conceptual modeling of research domains

1 Introduction

Curation and sharing research data to make it reusable and research result reproducible
is a topical issue over the years [13]. It is necessary since researchers use multiple
sources of open observational data and need to use the results of research in domain
communities. Heterogeneous data volumes as well as needs and directions to process
them grow, so research communities are not able to devote the most time to manual
resolution of data heterogeneity. Tools, formats, operation sets, and procedures shared
in research groups and communities are developed to maintain and perform research
data.

A collection of research workflows MyExperiment [14] includes registered
researchers joining interest groups and projects, using service libraries and shared
workflows. WF4Ever project [11] is aimed at preserving data, workflows and research
results for sharing and reuse. There is an awareness of the need to use global identification and metadata attributes with any resources. Workflows are accompanied by
everything required for their functioning. For this purpose, research objects (RO) are

© Springer Nature Switzerland AG 2019
Y. Manolopoulos and S. Stupnikov (Eds.): DAMDID/RCDL 2018, CCIS 1003, pp. 37–51, 2019.
https://doi.org/10.1007/978-3-030-23584-0_3

declared as containers that encapsulate data, metadata, workflows, documentation, links to external resources and share all resources related to research for a community. A set of operations for RO long-term preservation and access is defined. The wide community of researchers and data publishers FORCE11 [5] is focused improving the way of research communications by means of semantically-enhanced digital publications including links to data, software tools, mathematical models, protocols, and workflows.

In research data curation, efforts are intensified today to develop international, interdisciplinary research data infrastructures. Collaborative data infrastructures share various resources such as collections, archives, databases, storage and computing capacities, and provide services to search, access and manage them. EUDAT [19] is a network of numerous community-specific data repositories and some of Europe's largest data centers using common data services for data and service providers and research communities. EUDAT Collaborative Data Infrastructure (CDI) is a European infrastructure of integrated data services and resources to support research. Heterogeneous research data infrastructures interact to share research data globally and make science open.

European Open Science Cloud (EOSC) [2] initiative integrates research data repositories, sharing, long-term preservation access and reuse data across all disciplines. It supports full research lifecycle by providing access to software, services, protocols, methods from multiple disciplines and platforms. It makes obligatory Data Management Plans (DMP) which prescribes management and sharing data during a project. EOSC can coordinate and support federation of various data of data infrastructures in specific disciplines. The conception of Digital Objects (DO) [28] (partially similar to the RO conception in Wf4Ever) are containers representing data accompanied by persistent identifiers, metadata, code for processing and analysis the data. The Digital Object Access Protocol (DOAP) [29] defines standard operations to access and manipulate heterogeneous DOs using a common interface.

FAIR data principles [27] have gathered basic features used in data curation and preservation practices [13]. They are now being propagated in research data infrastructures and open science. These principles are aimed to provide data interoperability and reuse by machines and humans. For this purpose, data should be well identified, semantically defined with shared vocabularies and ontologies, accompanied by provenance information, comply with known protocols, standards, and data models, or have known mappings to them, and have clear access rules.

FAIR data principles have been defined informally. So they rise a variety of different interpretations from proposals of known technologies for providing FAIR principles, to simplified quantitative estimations and certification for conformity with FAIR principles [4, 12, 26]. So the authors of the principles had to explain what do and what don't FAIR data mean [18]. At the same time, it seems that FAIR data principles are sufficient to have some definite implications for requirements to research data infrastructures. Ones relevant to data semantics problems with respect to research communities are discussed in the paper. Having been applied at a semantically significant level, FAIR data principles are capable of changing research lifecycle in disciplinary communities and between them.

This investigation was presented shortly in [20], and more detailed conceptions are considered here. The rest of the paper is structured as follows. Today's research challenges are concerned in the next section. Section 3 tells about the role of formal specifications for data interoperability and reuse. Section 4 discusses approaches to data annotation and referencing in multidisciplinary infrastructures. In Sect. 5 a state of affairs in the astronomical domain and research in it as an example. Section 6 raises problems of correlation and method specifications of domain objects for research. The discussion of Sect. 7 is dedicated to an enhancement of research lifecycle using presented approaches of data specification in research communities.

2 Research on the Edge of Paradigms

Research in the paradigm of computational research included much code development and data integration efforts. As there are growing research communities and an increasing number of data sources, time spent by researchers in projects to data discovering and reusing is as high as about 80% [18]. A research lifecycle (Fig. 1) may include searching for data sources that may be heterogeneous and poorly documented, matching and integrating their data models and structures to make them accessible. Then the problem solving is implemented over those data sources in an internally developed data representation or in schemes of the data sources, and the resulting data are stored in a suitable format for preservation and usage in further work.

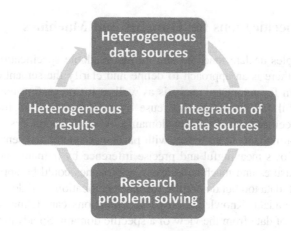

Fig. 1. A research approach in the computational paradigm

Some researchers continue using similar approaches to their research, although these approaches potentially bring multiple inefficiencies and additional work. Source reconciliation and integration problems are resolved for every new project. Resulting data become heterogeneous for other research groups. Non-trivial reuse of them leads to the necessity of data integration and development of the same methods multiple times. Scientific method implementations are tied to particular data sources. Programs are rewritten to solve the same problem with other or additional data sources.

Research processes using dynamically changed multiple data sources are shifting to the paradigm of data-intensive knowledge discovery [25]. Since FAIR data principles aim at making data interoperable and reusable for humans and machines, adherence to them should make reachable automation of data integration, linking data to methods, workflows and other related resources semantically relevant and applicable to them, and possibly automation of data-driven research process itself. For this purpose, metadata should describe data and services sufficiently for understanding their domain semantics and restrictions, their provenance and structure. On the other hand, they should be formal enough to allow reasoning by machines even without any human help. Relevant data and resources should not only be proposed to a human but used by machines in research process [18].

Research data infrastructures based on FAIR data principles in their turn should provide machine and human understandable metadata about data and services, and reasoning over metadata on all stages of data management. In multidisciplinary infrastructures, specifications of certain domains may be developed, collected and maintained by research communities working in them. Such specifications should continuously be reused in communities to reduce heterogeneities of multiple research data sources and to support cooperation of researchers and machines in communities and between them. Domain communities keep data FAIR by relating them to domain specifications.

3   Domain Specifications for Humans and Machines

FAIR data principles declare machine and human readable specifications of data. So data are FAIR if there is an approach to define and clarify the semantics of data in a knowledge domain for automated analysis as well as for presenting relevant data to a human. Meaningful data exchange and reuse by machines (helpful for humans too) require formal specifications of subject domains. A formal domain specification is a description of requirements to an object with particular syntax and semantics within a domain, which allows meaningful and precise inference by humans and machines.

Similarity measures and machine learning approaches could be applied for search and operating with data too but do not use formal specifications and do not provide sure inference over metadata. Knowledge-based specifications can define restrictions and permissible states of data from the view of a specific domain. So advanced ontological and rule-based models are preferable for metadata development.

Conceptualization and conceptual specifications are necessary not only in common and general research disciplines but in domains of interests of narrower and more specialized communities, as well as in overlapping domains, in which research problems and collaborations of research teams often occur. Most researches are held in an intersection of several domains, so they use constraints of several domains simultaneously to specify several points of view to the same objects of study. Reasoning in multidomain specifications should provide establishing relations and semantic interoperability between data belonging to different domains.

Particular syntax and semantics of domain specifications are achieved through the use of formal data models allowing automatic or interactive reasoning. Conceptual modeling of research domains includes developing ontologies and conceptual schemes. Ontologies define domain concepts and their relations. Conceptual schemes are abstract definitions of structure and behavior information objects in a domain.

Formal approaches to research domain objects specification keeping semantic interoperability between described objects are based on a kind of set inclusion or substituting (see Fig. 2). Non-formal approaches are not sufficient to provide such relations and to guarantee specification correctness and consistency.

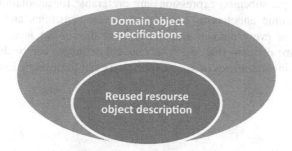

Fig. 2. Description of reused resources in terms of domain specifications

Some of the notions that may be used in data models for specification are:

- subclass relation for classes as sets of objects having common properties being considered;
- concept subsumption having formal interpretations in set theory for description logics [10] of various kinds allowing a subset of constructors;
- subtype relation including structural and behavioral specifications (Liskov) [16] and formal specification calculus (Kalinichenko) [15];
- specification refinement in B technology (Abrial) [9].

Semantic interoperability of data is accomplished through establishing relations between their specifications and mapping data with narrower specifications to a broader one. Interdisciplinary interoperability should be based on inclusion relations too. Development of global standards and using shared semantic specifications makes possible meaningful mapping and transfer data between domains through the mediation of them. Depending on the formality and granularity of specifications and standards, they provide different levels of reasoning. A general-level standard can only provide a general level of interoperability. For example, a standardized set of data operations does not allow to be sure of the relevance of the data transferred through the operations. It will require other means to be applied for formal control of data compatibility and reuse correctness.

4 Referencing Specification Approaches

Semantic annotation notion is used for a kind of metadata accompanying data or any other resources to semantic specifications such as ontologies. Most commonly used approaches of semantic annotation are based on informal references or instance-class relations, they use just binary relations.

To define the semantics of objects in interdisciplinary research annotations in terms of interrelated domain specifications may be required. So annotations as links to concepts from different domain specifications are insufficient, they do not express how an annotated object joins views from both domains. For multidomain specifications, subconcept (subtype, subclass) expressions are preferable for annotation models [23]. Such formal semantic annotations can express exact semantics and constraints of objects in terms of every domain and define how this object is constrained in an intersection of some domains (Fig. 3). A subconcept definition can be defined as a new concept over existing ontologies or as an expression even for a single instance that is an annotated object.

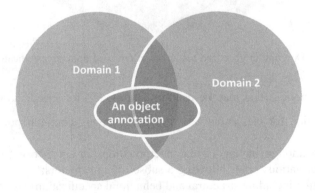

Fig. 3. Annotation in terms of several domains

Semantic annotations in a formal model allow reasoning over multiple domains using the same reasoning methods and instruments which are used for domain specification consistency verification. Evidence-based search for objects relevant to given one may be organized as inference of subconcepts or instances of subconcepts of the annotation expression.

5 The State of Art of Domain Specifications in Astronomy

An example of a discipline historically developing its information resources in a FAIR-like way is astronomy/astrophysics. Wide astronomical research community generates and uses many open data sources including sky surveys, catalogs, and databases of specific purposes. Data gathered at any time with any instrument remain valuable for analysis jointly with recent data. Virtual observatories providing access to available data sources are a basis of research in astronomy. International Virtual Observatory Alliance (IVOA) [6] includes national virtual observatories. It has developed formats, languages, tools, and repositories for accessing, sharing and querying astronomical data.

Data interoperability is supported by widely used standard formats such as FITS [3] for source images, VOTable [8] for tabular data. These formats encapsulate data and rich astronomical metadata. VOTable allows streaming in response to a query and supports separate metadata from data.

There are semantic specification standards, includes ontologies, thesauri, and conceptual models. Unified Content Descriptors (UCD) is the most commonly used specialized metadata controlled vocabulary. Unfortunately, it is not formal, has no well-defined semantics of annotations, and doesn't exclude description ambiguity and variance. Known astronomical ontologies are thesaurus-based and are not formal. For example, AstrObject ontology is a hierarchy of astronomical object names. Conceptual models (called data models in IVOA) are standardized structures in most common subdomains of astronomy, namely coordinate systems, photometry, spectroscopy, registered events, and others. They include necessary elements for objects and characteristics used in these subdomains. However, they are not usually used for data representation, but for data field annotations in some catalogs. ProvenanceDM is based on the W3C PROV-DM provenance model that defines relations between entities, actions, and agents. Observation Core Data Model (ObsCoreDM) [17] can be considered as an approach to the needs of domain specifications in astronomy. It joins UCDs, features of several standard conceptual schemes defining sky observation concepts, provenance model and allows querying them simultaneously.

Strasbourg Astronomical Data Center (CDS) [7] has built a digital repository that collects and shares astronomical data. The center has a data infrastructure with remotely accessible interfaces for data access, such as unified access to astronomical catalogs (VizieR), interactive atlas of the sky (Aladin), the database of astronomical objects (SIMBAD), specialized query language (ADQL), and the querying protocol (TAP) with access points. Although the interfaces are unified, catalog structures remain heterogeneity. They have human-readable metadata. Some common fields such as astronomical object coordinates and magnitudes are machine-readable and integratable into multisource queries. Various catalogs have heterogeneous formats of fields and value representations intractable for automatic processing. As a result, much time is spent on cleaning, value standardization and integration of data obtained with a seemingly unified interface.

ASTERICS project [1] as a part of research data infrastructure investigations of Horizon2020 program focuses on four large astronomical/astrophysical instruments generating data in different wavelengths. So the project develops tools for horizontally distributed and cloud-based repositories, processing data with workflows, access to analytical method libraries, matching methods for data from these instruments. It also integrates IVOA standards and services mentioned above to be used within the data infrastructure the Data Access, Discovery and Interoperability (DADI) package of the project. Multisource queries as a basis of interoperability allow joining objects by general data fields. However, packaging sundry services without implementing semantic-based approaches may cause that the shortcomings of the IVOA infrastructures can also be apparent in the data infrastructure developed in this project for more specific data fields and for various accessible data sources.

The experience of problem-solving within and among some research groups working in astronomy suggests the need to define well the research domain specifications and shows the advantages of their reuse with minimal extension for solving various research problems in this domain [21].

It required the development of a domain ontology and schemes for unifying data from heterogeneous sources. These specifications have been reused in solving a number of research problems.

There are modules of the ontology, which may be deemed as an upper ontology since they define concepts general not only for astronomy and used as a basis for the domain concepts. These domains are measurements, measurement quality, provenance, dependencies, events and processes, experiments. Domain ontology includes following general modules used in most astronomical research problems: astronomical objects, observation and instruments, astrometry, photometry, spectroscopy, stellar objects, astrophysical parameters, galaxies. More specific domains used as special research interests: binary and multiple stars, orbit movement, variable stars, light curves. More specific domain modules use concepts of generalized ones. All modules have mutual usage of concepts between them too.

Schemes of specialized subdomains have been created and reused as unified structures for mapping data sources to them and for solving research problems in terms of the same unified data presentations.

Data sources have been mapped to domain specifications, especially catalogs and surveys of photometry of single star (SDSS, 2MASS, USNOB-1, HD, HIP, and others), catalogs of visual binary stars (WDS, CCDM, TDSC), and different types of close binary stars (GCVS, SB9, ORB6, INT4). All of them have heterogeneous structures that are mapped into the same domain specifications. Once data sources mapped into domain specifications, they may be accessed in a unified structure accepted among the research groups using virtual or materialized view approaches.

Research activities have been carried out in overlapping domains in collaboration with several astronomical research groups using the same domain specifications. Analysis of multicolor photometry for different purposes, matching binary stars of various observational types from different catalogs, identification of high multiplicity hierarchical multiple stars, classification of eclipsing binary stars and other problems have been formulated using the same parts of specifications [21]. Common parts of the domain knowledge used in most problems are astrometrical coordinates and relative positions of stellar objects, and photometrical observations in particular photometrical systems.

Some problems use several specific subdomains together. For example, multiple stars may have pairs of components of different types having specific characteristics and constraints in their observation method domains. So to solve the problem of multiple star identification, pair identification rules have been derived from knowledge of different domains. Data on pairs of different types are acquired and transformed from multiple data sources mapped to domain specifications. Multiple star identifications have been represented as a new astronomical catalog [22].

6 Collections of Methods and Experiment Specifications

For comprehensive investigations of specific objects, it is important to share data about them, knowledge defining the semantics of entities and phenomena, the semantics of research methods applied to them, description of tools and workflows, publications, research results in that domain and other resources. No matter which kind of information object is used for research and shared for reuse, it should be supplied with metadata in terms of domain specifications. Any kind of information objects can be collected in repositories and described and categorized using formal metadata. Inference in formal metadata models makes it possible to select them from collections and access by selected global identifiers.

Data collections without collected methods related to these data make analysis and processing development difficult. So data packages are often linked to the method code for accessing or reproducing them. To provide automation of data processing or proposing existing method implementations to humans, a FAIR-like semantics-based approach implies not just linking but well-defined and consistent formal descriptions of data and methods to find and access them together. It means that methods applicable to research objects should be collected, considered as specific data kind and supplied with formal metadata for reuse. Method implementation collections may be defined as specifications of correlation of research object characteristics and implement relations between them [24].

In research data infrastructures, methods used in any research domain should be conceptually specified and collected in addition to general-purpose methods such as multidimensional data analysis or machine learning. For example, it is insufficient to say that a machine learning method is applied to research object data. Any application of a machine learning method has its physical sense in the domain. Thus, to apply a machine learning method to an object in the domain, a researcher should make the following:

- wrap a machine learning method with a domain object method definition having specific domain semantics for this object;
- define metadata of the method to describe its semantics in the domain;
- define metadata describing preconditions related to inputs and postconditions related to outputs of the method in terms of the domain to restrict data applicable with the method;
- define provenance metadata describing that the method has been implemented with specific machine learning method application.

Meaningful access to known implementations of research object methods should be provided to humans and machines by a query. Queries can specify the semantics of methods as a type of object correlation in the domain, known and evaluated parameters, mathematical approaches to method implementations, authoring, policies, and other condition.

Experiments over data in research infrastructures are constructed using shared and interoperable data, methods (services), and processes (workflows). Research experiments can include data analysis, modeling in accordance with hypotheses and testing models by observational data. Besides providing access to data and method implementation collections, research infrastructures should include workflow metadata, workflow collections, and instruments for experiment supporting hypothesis generating and testing.

Specifications of workflows include formal annotations of workflows as wholes and annotation of their elements. Specifications of elements, their inputs and outputs in terms of the domain allow linkage correctness verification, automation of workflow development, reuse of workflows as wholes or fragments of workflows.

Research experiment support includes manual or automated hypothesis generation and testing. Correlation concept used for method specifications could also be applied to specify hypotheses as estimated correlations between investigated object characteristics and expected reflections of values of other characteristics. So hypothesis specifications draw on the structure of the research object in domain specifications.

Models are implemented as subconcepts in accordance with the hypothesis structure. Modeled data are generated and represented as instances of model specifications.

The same domain object correlation specifications are used to test hypotheses by observed object data. Relevant data sources meeting requirement of the domain could be found in data collections. Differences between generated and observed data are analyzed to test hypotheses [24].

Research results may include gathered, reformatted, selected, modeled, derived data, metadata including semantic annotations and provenance, new domain specifications, implemented methods, programs, workflows, mathematical models, hardware calculation and preservation resources, software tools, and others. All these resources may be shared in some policy.

7   Research Lifecycle in Domain Communities

Since shared semantics of research objects are becoming increasingly important for data reuse in each discipline or subject domain, communities working in a domain should have conceptual specifications related to their research and maintain a strong commitment to them.

Communities of researchers and vendors of analytical tools and research instruments and data owners are interested in the long-term shared access to heterogeneous data and method collections. A natural way of conceptualization and formal specification of a domain is the development of them in communities stimulated by a necessity to reach semantic interoperability of interacting components, integration of data collections, reuse of data resources and method reproducibility due to binding to the semantics of the subject domains.

Joining of humans and machines in a community means that they operate within the ontological commitment defined by its shared ontologies, i.e. use of the concepts of the subject domain in a consistent way with respect to the theories specified by the ontologies. Ontologies are important for the automation of consistency control on any manipulations with the domain concepts.

An interaction of communities for solving interdisciplinary problems requires simultaneous querying using different domain vocabularies. In that case, the researchers should commit to the specifications of several domains. Research communities may be built hierarchically. More specific research interests are based on standards of a broader community and are specified if not covered in it.

Development of domain specifications takes a lot of work for domain analysis and agreement, however research communities constantly working in a domain can be interested in formal specifications of their domains. Research communities develop ontologies and most common conceptual schemes for their domains, integrate related data sources, collect implemented methods. These efforts are one-time investments in building essential resources that are later continuously reused.

Maintenance of shared domain specifications becomes a basis for arranging collections of data and sources, collections of specific methods, embedding research results into such collections for further research.

A formal approach to metadata maintenance and research process allows defining a set of typical operations that research communities will use to solve their problems, to analyze data in the domain and to curate data and metadata. They all access or modify domain specifications, and they can have unified interfaces independent of the chosen formalism of specification inclusion or substitution. These operations should be implemented using means chosen formalism. Operations can include:

- exploring domain specifications;
- manipulations for building domain specifications with obligatory consistency verification of resulting specifications;
- reasoning interdomain subconcept relations between specifications;
- annotating various resources with subconcept expressions in terms of multiple domain specifications;
- registering schemes and implementations of methods in collections with annotations;
- searching for and accessing relevant data whose annotations are subconcepts of queries;
- searching for and applying relevant methods, services, workflows or workflow fragments whose semantics are defined in annotations as subconcepts of queries, preconditions are strengthened by input data, and postconditions weakened by required outputs;
- perhaps automizing entire research process using domain specifications, problem specifications, accessible data, and methods.

Since data sources and collections are integrated using domain specifications, and method implementations are linked to domain object specifications, research problems are solved using relevant resources from collections. From all the sections above, activities of communities are defined by research lifecycle (Fig. 4) to provide data interoperability and reuse over domain specifications.

1. A research problem is described by researches (or generated by machines) as a specification of requirements in terms of domain specifications. It means that domain object structures, knowledge constraints, method signatures are used to express the demands as object restrictions, specific relations, or process specifications as sequences of requirement specifications to solve the problem.

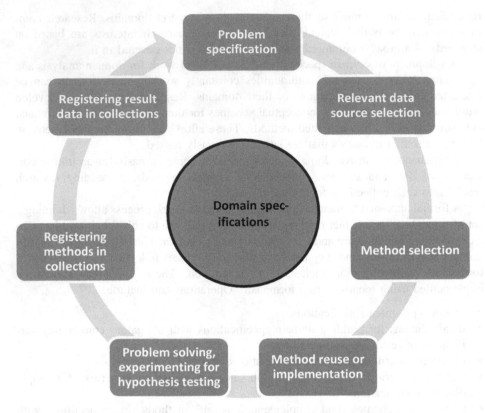

Fig. 4. A research life cycle on the basement of domain specifications.

2. Data relevant to specifications of requirements are found in data sources and repositories by semantic annotations as subconcepts of requirements. Data is selected manually or automatically from accessible relevant one to be reused in problem-solving. If found data is insufficient, an observation plan and data acquisition may be run. New data should be annotated in terms of domain specifications and have provenance information hoe they have been collected.

3. Relevant methods implementations are found and selected from method/service/ process collections by semantic annotations of their semantics and constraints of their inputs and outputs. Missing methods should be implemented or developed as a workflow. New implemented methods should be annotated in terms of domain specifications in accordance with the requirements they meet. They should be specified with provenance information about authors, way of implementation, and others.

4. The problem is solved with the reuse of selected relevant for this purpose. The process of problem-solving may be controlled manually or automatically by queries, method calls, programs, workflows, experiments plans. Hypotheses may be tested with relevant observational data by comparison with modeled data. The results of problem-solving can include new knowledge and models, modeled and

calculated data, data slices and classifications, new implemented methods and workflows, specifications of new subdomains, and other information. These result should be defined with semantic annotations in terms of domain specifications. New data should be specified with provenance information on authors, from which data and with which methods they have been obtained or derived.

5. Methods and workflows implemented for some requirements in the domain may have some sharing policy with a group of researches, community or for open reuse. For this purpose, they and registered in method collections with metadata annotations in terms of the domain specifications.

6. New resulting data may be preserved in repositories, registered in collections with metadata describing them for reusability.

The proposed life cycle of a research process based on the formal domain specifications meets the challenges of a new research paradigm and FAIR data principles, provides semantic interaction within research communities, avoids repeated activities for data harmonizing and integrating, and developing methods implemented earlier. Development of domain specifications and data source integration process are performed once and reused in communities. Research problems are formulated in terms of domain specifications and can reuse any relevant data and methods registered in collections within a data infrastructure.

8 Conclusion

FAIR data principles have been informally defined but have ambitious requirements of machine- and human-readable data and metadata. In research infrastructures, this expectation leads to formal reasoning in the metadata. So we have traced how these requirements may be reflected in the usage of formal domain specifications of research objects and in the automation of research processes including work with problem specifications, method specifying and collecting, and holding research experiments. A key role of communities in the development and maintenance of formal domain specifications has been shown.

Acknowledgments. The work was supported by the Russian Foundation for Basic Research (grants 18-07-01434, 18-29-22096, 19-07-01198).

References

1. ASTERICS: Astronomy ESFRI & Research Infrastructure Cluster. https://www.asterics2020.eu/. Accessed 01 Jan 2019
2. EOSC Declaration. https://ec.europa.eu/research/openscience/pdf/eosc_declaration.pdf. Accessed 01 Jan 2019
3. FITS: Flexible Image Transport Specification. http://fits.gsfc.nasa.gov/
4. Guidelines on FAIR Data Management in Horizon 2020. Directorate-General for Research and Innovation European Commission (2016). http://ec.europa.eu/research/participants/data/ref/h2020/grants_manual/hi/oa_pilot/h2020-hi-oa-datamgt_en.pdf. Accessed 01 Jan 2019

5. Improving Future Research Communication and e-Scholarship. Bournea, P., Clarkb, T., Dalec, R., de Waardd, A., Hermane, I., Hovyf, E., Shotton, D. (eds.) The Future of Research Communications and e-Scholarship (2011). https://www.force11.org/. Accessed 01 Jan 2019

6. International Virtual Observatory Alliance. http://www.ivoa.net

7. Strasbourg Astronomical Data Center (CDS). http://cdsportal.u-strasbg.fr/

8. VOTable Format Definition. Version 1.3. IVOA Recommendation. IVOA (2013). http://www.ivoa.net/Documents/latest/VOT.html. Accessed 01 Jan 2019

9. Abrial, J.-R.: The B-Book: Assigning Programs to Meanings. Cambridge University Press, Cambridge (1996)

10. Baader, F., Horrocks, I., Lutz, C., Sattler, U.: Introduction to Description Logic. Cambridge University Press, Cambridge (2017)

11. Belhajjame K., et al.: Workflow-centric research objects: a first class citizen in the scholarly discourse. In: ESWC2012 Workshop on the Future of Scholarly Communication in the Semantic Web (SePublica2012), Heraklion, pp. 1–12 (2012)

12. Doorn, P., Dillo, I.: FAIR Data in Trustworthy Data Repositories. DANS/ EUDAT/ OpenAIRE Webinar (2016). https://eudat.eu/events/webinar/fair-data-in-trustworthy-data-repositories-webinar. Accessed 01 Jan 2019

13. Hodge, G.M.: Best practices for digital archiving: an information life cycle approach. D-Lib Mag. 6(1) (2000). ISSN 1082-9873. http://www.dlib.org/dlib/january00/01hodge.html. Accessed 01 Jan 2019

14. Goble, C.A., De Roure, D.C.: myExperiment: social networking for workflow-using e-scientists. In: Workflows in Support of Large-Scale Science, pp. 1–2. ACM (2007)

15. Kalinichenko, L.A.: Compositional specification calculus for information systems development. In: Eder, J., Rozman, I., Welzer, T. (eds.) Advances in Databases and Information Systems. LNCS, vol. 1691, pp. 317–331. Springer, Heidelberg (1999). https://doi.org/10.1007/3-540-48252-0_24

16. Liskov, B., Wing, J.: A behavioral notion of subtyping. ACM Trans. Program. Lang. Syst. (TOPLAS) 16(6), 1811–1841 (1994)

17. Louys, M., et al.: Observation data model core components and its implementation in the table access protocol. Version 1.1. IVOA Recommendation, 09 May 2017. IVOA (2017). http://www.ivoa.net/documents/ObsCore/. Accessed 01 Jan 2019

18. Mons, B., et al.: Cloudy, increasingly FAIR; revisiting the FAIR data guiding principles for the European open science cloud. Inform. Serv. Use 37(1), 49–56 (2017). https://doi.org/10.3233/isu-170824

19. Schentz, H., le Franc, Y.: Building a semantic repository using B2SHARE. In: EUDAT 3rd Conference (2014)

20. Skvortsov, N.A.: Meaningful data interoperability and reuse among heterogeneous scientific communities. In: Kalinichenko, L., Manolopoulos, Y., Stupnikov, S., Skvortsov, N., Sukhomlin, V. (eds.) Selected Papers of the XX International Conference on Data Analytics and Management in Data Intensive Domains (DAMDID/RCDL 2018), vol. 2277, pp. 14–15. CEUR (2018). http://ceur-ws.org/Vol-2277/paper05.pdf. Accessed 01 Jan 2019

21. Skvortsov, N.A., Avvakumova, E.A., Bryukhov, D.O., et al.: Conceptual approach to astronomical problems. Astrophys. Bull. 71(1), 114–124 (2016). https://doi.org/10.1134/S1990341316010120

22. Skvortsov, N.A., Kalinichenko, L.A., Karchevsky, A.V., Kovaleva, D.A., Malkov, O.Y.: Matching and verification of multiple stellar systems in the identification list of binaries. In: Kalinichenko, L., Manolopoulos, Y., Malkov, O., Skvortsov, N., Stupnikov, S., Sukhomlin, V. (eds.) Data Analytics and Management in Data Intensive Domains. DAMDID/RCDL 2017. Communications in Computer and Information Science, vol. 822, pp. 102–112. Springer, Heidelberg (2018). https://doi.org/10.1007/978-3-319-96553-6_8

23. Skvortsov, N.A., Vovchenko, A.E., Kalinichenko, L.A., Kovalev, D.A., Stupnikov S.A.: Metadata model for semantic search for rule-based workflow implementations. In: Systems and Means of Informatics. vol. 24, Iss. 4, pp. 4–28, IPI RAS, Moscow (2014). (In Russian)

24. Skvortsov, N.A., Kalinichenko, L.A., Kovalev, D.A.: Conceptualization of methods and experiments in data intensive research domains. In: Kalinichenko, L., Kuznetsov, S., Manolopoulos, Y. (eds.) Data Analytics and Management in Data Intensive Domains (DAMDID/RCDL 2016). CCIS, vol. 706, pp. 3–17. Springer, Cham (2017). https://doi.org/10.1007/978-3-319-57135-5_1

25. Tolle, K.M., Tansley, D.S.W., Hey, A.J.G.: The Fourth paradigm: data-intensive scientific discovery [point of view]. Proc. IEEE. **99**(8), 1334–1337 (2011). https://doi.org/10.1109/jproc.2011.2155130

26. Wilkinson, M., et al.: Interoperability and FAIRness through a novel combination of web technologies. PeerJ Preprints **5**, e2522v2 (2017). https://doi.org/10.7287/peerj.preprints.2522v2

27. Wilkinson, M., et al.: The FAIR guiding principles for scientific data management and stewardship. Sci. Data **3**, 160018 (2016). https://doi.org/10.1038/sdata.2016.18

28. Wittenburg, P.: From persistent identifiers to digital objects to make data science more efficient. Data Intell. **1**(1), 6–21 (2019). https://doi.org/10.1162/dint_a_00004

29. Wittenburg, P., Strawn, G.: Common Patterns in Revolutionary Infrastructures and data. RDA (2018). https://www.rd-alliance.org/sites/default/files/Common_Patterns_in_Revolutionising_Infrastructures-final.pdf. Accessed 01 Jan 2019

Knowledge Representation

Tabular and Graphic Resources
in Quantitative Spectroscopy

Nikolai A. Lavrentiev, Alexey I. Privezentsev,
and Alexander Z. Fazliev$^{(\boxtimes)}$ (iD)

Institute of Atmospheric Optics SB RAS,
V.E. Zuev Sq.1, Tomsk 634055, Russia
{lnick, remake, faz}@iao.ru

Abstract. An approach to forming applied ontologies in subject domains in which data are presented in various forms of tables and scientific graphics is proposed. A description of the sources of data and information presented in this form is given. Using quantitative spectroscopy as an example, an approach to forming semantic annotations characterizing these sources is demonstrated. The major types of the sources are described. For scientific graphics, an approach to solving the problem of reducing and systematizing the graphic resources to search for plots in the subject domain is described. A partition into groups of functions used in the plots that are not interrelated with each other is constructed to define different spectral functions to be equivalent. The metrics of three applied ontologies of spectroscopy used in comparing data collections are briefly described.

Keywords: Systematization of large data · Quantitative spectroscopy · Applied ontologies

1 Introduction

In information resources related to subject areas with intensive use of data, numerous research results are presented in tabular and graphic forms. As a rule, in search for information this part of resources is ignored. The processing of such resources is considered to be not economically feasible due to the lack of universal software for detailed description of such resources from various subject domains.

Back in the 90s search for information in tabular and graphic resources was made using metadata integrated into html-pages. In the early 2000s [1] Semantic Web technologies were created to replace the traditional metadata by semantic annotations. The transition to semantic annotations when describing scientific information resources was not complete, since, on the one hand, the application of the new technologies turned out to be a difficult process and, on the other hand, there was no demand for detailed queries that give unambiguous answers.

At the initial stage of creating the Web, the non-scientific resources greatly exceeded the scientific resources. The situation changed at the end of the 2000 s when the scientific data volume began to increase dramatically [2, 3]. The scientific information resources are available on the Internet in the form of publications (files), data collections

© Springer Nature Switzerland AG 2019
Y. Manolopoulos and S. Stupnikov (Eds.): DAMDID/RCDL 2018, CCIS 1003, pp. 55–69, 2019.
https://doi.org/10.1007/978-3-030-23584-0_4

(databases), ontologies of subject domains (knowledge bases), etc. We will consider mainly tabular and graphic presentation of data in scientific articles and their systematization. These resources are chosen, on the one hand, since they have been traditionally used in research and, on the other hand, there exists a need to search for resources in graphic form, with a high degree of detail of the queries. In the mid-2000 s there were some attempts to systematize non-textual parts of scientific resources in some subject areas [MPI, our works]. In the present paper, details of the systematization methods will be demonstrated by using examples from quantitative spectroscopy.

Over the past 15 years, the authors have been systematizing spectral data sets on spectroscopy. The semantic annotations of such data sets resulting from this work have become part of applied ontologies, in particular, characteristics of one of the main properties of such sets – the trust of these data [4]. The tables and plots representing the parameters of spectral lines and spectral functions have been digitized. The tables were digitized to control the quality of the published measurement data, and the spectral functions were digitized to obtain spectral information when there are no accurate results, as well as to control the asymptotic behavior of calculation data.

We have constructed applied ontologies characterizing the quality of information resources on molecular spectroscopy [5], states and transitions of atmospheric molecules [6], and ontologies of graphic resources on spectroscopy [7].

These ontologies describe the properties of tabular data characterizing spectral lines, which have been well studied over the past 80 years. In the first thirty years of this period, the publications contained, along with small data tables, a considerable number of scientific plots describing spectral functions. The creation of Fourier spectrometers in the late 60s initiated the appearance of large numerical arrays of accurate data on the parameters of spectral lines, and in the subsequent years the graphic representation of spectral data in high-resolution quantitative spectroscopy was replaced by a tabular presentation.

Nevertheless, spectroscopy still has areas in which it is difficult to achieve high accuracy of the spectral characteristics by using the modern experimental equipment. These areas include continuum absorption, which is important in studying planetary and exoplanetary atmospheres [8, 9] and the spectral properties of weakly bound molecular complexes and molecules in the UV range needed for quantitative description of the photochemical reactions in the gas phase [10]. In these subject areas, the volume of the spectral information contained in scientific graphics is much greater than that of the information available in tabular form.

The main idea in systematizing scientific graphics is in isolating, in composite plots containing several sets of curves, every curve of this set in a separate primitive plot. Such a plot is provided with an additional set of metadata describing it with the detail needed when searching.

This paper represents the extended version of our report published in the proceedings of the DAMDID/RCDL 2018 conference [11]. It is significantly expanded by focusing on the description of scientific graphics. First of all, the definitions of different types of scientific plots are clarified, and the classification of the research plots and figures is described. The classification of spectral functions and their arguments is briefly characterized. The semantic heterogeneity of terms, used in descriptions of spectral functions and arguments, is discussed. The conclusion is completely rewritten, and some fragments of introduction are corrected.

2 Some Peculiarities of Tabular and Graphic Presentations of Resources in Publications

2.1 Publication Model

Scientific publications are most often used to store, transmit, and analyze the information contained in them. Traditionally scientific articles contain a text in a natural language with mathematical equations, chemical reactions, physical formulas, tables, plots, figures, etc. The text is mostly used to search for the information requested by the user. In many subject areas, numerical arrays are used in tabular and graphic forms; these are solutions of computational problems, measurements, or observations. Each of the solutions is part of a publication containing a large number of typical facts. Equations, formulas, and sets of reactions are more abstract resources, because most of these resources do not have unique names, and any way to annotate them requires a certain level of professional training.

In a simple case, to form semantic annotations characterizing tabular and graphic publication resources one can only describe the properties of solutions in the subject domain used in such presentations. The solution of a computational problem is a numerical array, which, being supplemented by a set of its properties, can serve as a more accurate formal model of those publication parts in which these arrays are presented in tabular or graphic form. The specification of a set of properties is determined by search tasks that are of interest to the researchers in a given subject area and consumers of these numerical arrays in applied subject areas.

The choice of a publication model for sets of numerical arrays in tabular and graphic forms is associated with automatic cataloging of such information resources in a subject domain. The authors' collection of articles in the field of quantitative spectroscopy exceeds 12,000 publications from 1886 to the present. The chosen subject domain model [12] contains solutions of seven spectroscopic problems which are important for applied subject domain, such as astronomy, atmospheric optics, spectroscopy, etc.

In publications, tables may contain not only numerical arrays, but also scientific graphics. The graphic resources in scientific subject domain can be divided into two parts: typically 2- and 3-D mathematical plots and other images. Now digital images of scientific graphics can be found in appendices of some journals to make quantitative comparisons of graphics at lower cost.

2.2 Tabular Presentation

Wide use of numerical data has brought a great variety of forms of tabular presentation. Table data in articles and plain text files contain numerical arrays with positional formatting using whitespace characters or enumeration formatting with separating characters, for example, CSV (comma-separated values) files. The subject area model chosen by us in the W@DIS information system [12] makes it possible to describe the intension structure of semantically significant numerical arrays in tabular form. When systematizing tabular data, the form and structure of a table are not described. Thus, not all information published in tabular form is semantically annotated, but only that necessary for studying the quality of tabular data (Fig. 1).

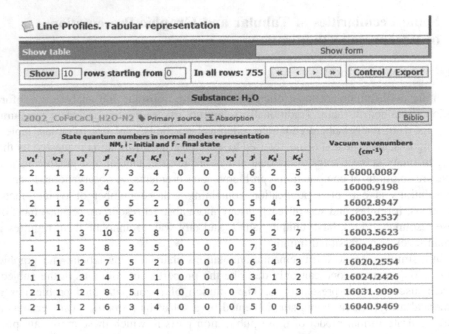

Fig. 1. Typical table representation in the W@DIS information system as exemplified by the numerical array.

In the W@DIS information system described below, the numerical arrays from tables contained in the articles are major resources.

In the future, the volumes of published numerical arrays will increase; the arrays will be published only in the appendices, and the text part of a publication will include an analysis of computer-generated semantic annotations.

2.3 Scientific Graphics

Simplified Conceptualization of Scientific Graphics. In quantitative spectroscopy, scientific plots are used where there are no precise measurements by modern experimental equipment (for example, in case of complex structures of molecules or complexes or when the spectral contributions of the components of mixtures or spectral measurements in the short-wave radiation range cannot be separated). Scientific graphics is still used in the field of continuum absorption, planetary and exoplanetary atmospheres and the spectral properties of weakly bound molecular complexes and molecules in the UV range to quantitatively describe the rates of photochemical reactions in the gas phase. The main role of such plots is in comparing data arrays to identify qualitative features (proximity, equivalence, similarity, etc.) of these data.

Scientific graphics may be divided into two parts. In spectroscopy, the scientific graphics constructed using algorithms is mostly represented by mathematical plots. Below they will be called scientific plots or simply plots. The plots available in publications may be divided into two classes: primitive and composite ones.

Primitive plots have a single system of coordinates with a single numerical array having the form of a curve, a set of points or sticks. Primitive plots have some properties that are important for classifying them. One of these properties is the type of numerical array used to construct the curve. This property has two meanings: of an original array (that is, an array obtained by the authors of the publication) or of a cited one (that is, an array published earlier). Thus, all primitive plots are divided into two groups: original plots and cited plots. Unlike primitive plots, composite plots may contain multiple curves (sets of points or sticks) in one system of coordinates. The published composite plots may also be divided into two groups: original and cited ones. The first type includes plots containing only the original curves of the authors of the publication being considered, and the second one, plots containing both the original and cited curves.

Plots in a publication are part of figures which may contain primitive plots, composite plots, and sets of primitive and composite plots. Figures may be primitive or composite. A primitive figure may be a primitive or a composite plot. A composite figure contains at least two plots.

The second part of scientific graphics is graphics that cannot be presented in articles in a unified way or such a presentation is difficult to realize. For the elements of such graphics, we will use the term "images". Examples of such elements are images of surfaces in raster graphics.

The scientific graphics described in this paper represents functions of physical quantities in (1D-2D) Cartesian systems of coordinates. 2D plots are most widely used. As a rule, a plot in a system of coordinates consists of several curves characterizing the behavior of physical quantities under various thermodynamic conditions, for example, describing a comparison of the original results of the authors with those obtained in works of other researchers. The number of plots containing only one curve in the total volume of published plots is relatively small.

Definition of Concepts in the GrafOnto System. After describing the conceptual components of the subject area, we give several definitions:

Definition 1. A *primitive plot* extracted from a published figure is that containing one curve in the same system of coordinates as in the figure, with the same physical quantities, units of measurement, and set of metadata.

There are two types of primitive plots in the GrafOnto system (http://wadis.saga.iao.ru/complexes): original and cited ones. Cited plots are those explicitly cited by the authors in the publication. Otherwise primitive plots are considered original.

Definition 2. A *composite plot* is a plot in Cartesian coordinates containing all *primitive plots* (more than one) from a published figure and a set of metadata describing the composite figure.

The GrafOnto system has three types of composite plots: original, cited, and multi-paper ones. An original composite plot contains only original primitive plots. A cited composite plot contains at least one cited primitive plot. A multi-paper plot contains at least one original or cited plot and at least one original primitive plot corresponding to one of the cited ones.

Definition 3. A *primitive image* in a published figure is an image of an object under study that does not have an adequate mathematical description and a set of metadata characterizing its properties associated with the object.

Definition 4. A *composite image* in a published figure is an image containing more than one primitive image.

In particular, the set of metadata of a primitive image includes a bibliographic reference to the publication from which the figure is extracted. Composite images can be one- or multi-paper.

Definition 5. A *primitive figure* is a figure containing one scientific plot or image.

Definition 6. A *composite figure* is a figure containing several scientific plots and/or images.

Below Figs. 2 and 3 represent the composite plot and figure from Refs. [13, 14], respectively.

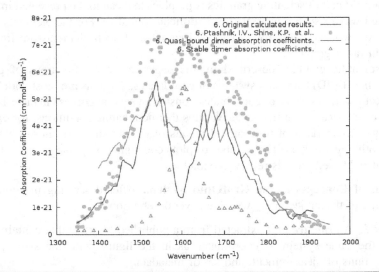

Figure 6. Experimental H2O self-continuum (T =296 K) versus predicted absorption spectra of bound and metastable dimers1 and present calculations for the 1400–1900 cm-1 band. The black dots are the experimental data, the triangles are the stable dimer absorption coefficients, the black curves are the quasi-bound dimer absorption coefficients, and the gray curves represent the calculated results using the κLorγ line shape. The frequency step is 50 cm-1.

Reference
Olga B. Rodimova,
Continuum water vapor absorption in the 4000–8000 cm⁻¹ region,
Proc. SPIE 9680, 21st International Symposium Atmospheric and Ocean Optics: Atmospheric Physics,
SPIE - The international society for optical engineering, **2015**, Pages 968002,
DOI: 10.1117/12.2205332.
Annotation

Fig. 2. An instance of composite plot equivalent to Fig. 6 from Ref. [13]

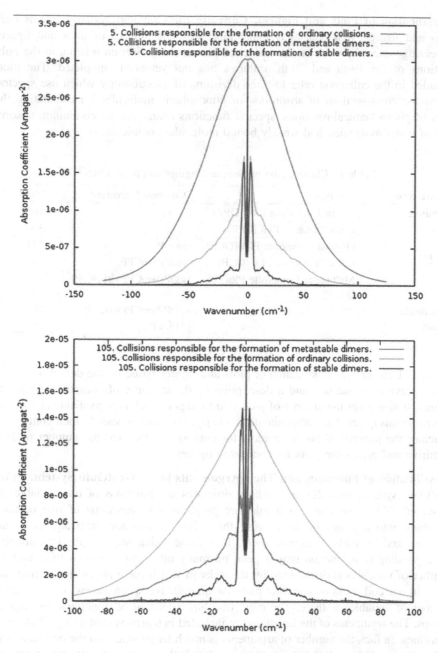

Figure 5. Spectral function components for the CO$_2$–Ar (a) and CO$_2$–Xe systems (b) at 296 K. The curves show the contributions from collisions responsible for the formation c stable (1) and metastable (2) dimers and ordinary collisions (3).

Fig. 3. An instance of composite figure constructed with the data provided by the authors of Fig. 5 [14].

Classification of Plots and Figures. Currently the GrafOnto system contains only plots and figures; it is constantly expanding, with the number of plots and figures increasing, since the work on the historical part of the collection relating to the publications of the 19-th and 20-th centuries has not yet been completed. The plots included in the collection refer to three divisions of spectroscopy which use spectral functions: cross-sections of absorption of atmospheric molecules for calculating the rates of photochemical reactions, spectral functions characterizing continuum absorption of water molecules, and weakly bound molecular complexes.

Table 1. Classification of plots and figures and their structure

Figure type	Plot type	Computed structure	Statistics
Primitive Figure	Original Primitive Plot (OPP)		1634
	Cited Primitive Plot (CPP)		370
	Original Composite Plot (OCP)	nOPP	331
	Cited Composite Plot (CCP)	nOPP+mCPP	131
	Multipaper Composite Plot (MCP)	nOPP+mCPP+k{OCiP}	
Composite Figure		nOPP+mCPP+kOCP +pCCP	76

Table 1 presents a classification of plots and figures based on the definitions given in the previous subsection and a description of the structure of composite plots and figures. It also gives the number of plots and figures of each type available at the time of writing this paper. The composite plots and figures structure specify their contents. It indicates the number of primitive plots for composite plots and the number of both primitive and composite plots for composite figures.

Classification of Functions and Their Arguments in the GrafOnto System. In the GrafOnto system, only 2D plots whose ordinates are functions of one variable are presented. These functions, as a rule, are projections of functions of two or three variables onto a plane. In most cases, the values of functions and arguments are characterized by units of measurement. In a subject domain, the physical quantity corresponding to a function may depend on many other physical quantities, and the number of variables is determined by the tasks in which these physical quantities are used. In our collection of scientific plots for quantitative spectroscopy, the maximum number of variables is three. The collection has a total of 69 functions forming 11 groups. The arguments of the functions are included in 6 groups containing 22 physical quantities. In fact, the number of arguments is much larger since, on the one hand, our collection is not complete but, on the other hand, the variety of units for measuring length, area, volume, frequency, and density is much greater than that shown in the plots of our collection. Table 2 shows a classification of the functions by their arguments.

Table 2. Classification of functions and their arguments in the GrafOnto system

	Frequency (12)	Excitation Energy(1)	Temperature, Pressure (4)	Cluster size (1)	Density (2)	Distance (2)
$S_1(37)$	x					
$S_2(1)$	x	x				
$S_3(1)$	x	x	x			
$S_4(7)$	x		x			
$S_5(1)$	x			x		
$S_6(1)$	x		x	x		
$S_7(1)$	x				x	
T (9)			x			
C(1)				x		
D(1)					x	
P(2)						x

Synonyms of the Names of Functions and Arguments in Scientific Graphics. The problem of semantic heterogeneity is typical for almost all subject domains in which the conceptualization and models are created by various scientific groups of researchers. As in other methods of presenting information in science, semantic heterogeneity in scientific graphics is due to differences in the educational and cultural level of the researchers, lack of an established terminology, and the use of simplifications. These simplifications are caused by the replacement of the spelling of the terms and concepts by their letter or symbolic surrogates. Let us give some examples from the practice of avoiding the semantic uncertainty of functions and their arguments in the GrafOnto system.

In our collection, the physical quantities most used in the plots are the argument Wavenumber (cm^{-1}) [Frequency (cm^{-1}); IR Energy (cm^{-1}); Photon Energy (cm^{-1}); $\sigma(cm^{-1})$; ν, cm^{-1}; $1/\lambda$ (cm^{-1}); $\sigma(cm^{-1})$; ω, (cm^{-1}), etc.] and the function Absorption Coefficient $(cm^2\ molecules^{-1}atm^{-1})$ [Self-continuum cross-section; Cs Self-broadening Coefficient; $C_s(T)$; C_N^0; C_s; $\alpha(\omega)$, C_{self}; k; C_S^0; $-\ln C_S^0$; C_f, etc.]. Examples of synonyms and simplifications of terms and concepts are given in the square brackets. In most cases, the semantic uncertainty of terms and concepts can be avoided if the units of measurement are explicitly given for physical quantities.

In the GrafOnto system, synonyms and surrogates for the terms and concepts are one of the parts of the collection of plots. They are used in solving the reduction problem, when equivalence is established between individuals and classes that are clearly defined in ontology and their analogs with surrogate names.

3 Information and Data Sources

3.1 Definitions

The molecules, for which the spectroscopy problems mentioned in [11] have been solved, number in the thousands. These problems were solved using different methods (experimental, theoretical, etc). Some of their solutions uniquely defined the states and transitions of molecules. Below we discuss only the solution of such kind. For this reason, solutions to several problems by different methods for different molecules or their isotopologues can be presented in one publication. The solution to one task can be the content of several tables. During systematization of data extracted from publications, such a variety of tables creates many problems, especially in the cases where the solution to a subject task is divided into parts and is represented in several tables. There is no sense to refer individual numerical arrays extracted from tables of a publication to these tables. For the purpose of integrating different parts of the solutions we use an information object that represents the original data of a publication relating to one molecule, one spectroscopy task, and one solution method. One can use two different property sets to define research plots. The first set includes the properties of the numerical array, used for the construction of the plot and the second one contains the geometric properties of the curves, points or sticks. These two sets complement each other. Our paper concentrates on the first set of properties.

Primitive and Composite Data Sources. This information object shall be called the data source. Different data source types are met in scientific papers. Let us give several definitions.

Definition 7. All parts of the published solution to a task of quantitative spectroscopy along with the molecule name, reference, and name of the solution method (or reference to the method description) are called the "primitive data source".

Note that this definition does not depend on how the numerical array is represented in the publication. We assume that empty solutions are not published. On the other hand, solutions can include measurement data which go out of date with time or wrong solutions. The data source, containing the problem solution completely declined by experts, is called negligible. The number of such sources in the modern spectroscopy is insignificant. It's worth noting, that curves from the plots are not clearly defined by numerical arrays. Such kind of array is created by digitizing a corresponding curve; therefore the downside of such arrays is low precision. During the collection formation stage this problem was solved by contacting the publications' authors and asking them to provide original numerical arrays.

Definition 8. An information object exhibiting basic properties of a primary source of data cardinality of which differs from unity is called the composite data source.

Note that a composite data source should uniquely define included states and/or transitions. Any expert set of spectral data (e.g., [15–18]) can serve an example of composite data source.

Information Source. A primitive data source can be endowed with additional properties. The list and number of these properties depend on information tasks for solution of which these properties are used. A data source with additional properties is called the source of information.

Definition 9. A primitive (composite) data source with additional properties is called a primitive (composite) source of information extracted from a publication.

The source of information is a set of properties and their values attributed to a data source. For a number of information tasks, for example, the search for reliable solutions to quantitative spectroscopy problems, one can select properties values of which are automatically calculated. A source of information usually includes some statements from the publication that contains the data source described by this source of information. The better half of a source of information characterizes the knowledge contained in the publication in an implicit form.

The list of additional properties is determined by a researcher on the basis of information tasks that are to be solved. There are two such tasks in our work: the task of semantic search of the most fitting information sources and the task of automated composition of an expert composite information source in quantitative spectroscopy [19, 20]. Let us note that primitive sources of information relating to one publication do not contain identical statements. The difference between a publication and a related composite information source can be significantly smaller than the difference between the publication and a related primitive data source. This is due to those additional properties of the task solution in the publication that are included in the definition of a particular source of information. For example, such an additional property can be the description of trust of the solution or the description of the standard deviations of the initial data source from other data sources, etc. In addition, the statements contained in the primitive or composite source of information may not be contained in the publication.

An example of brief information source description is given in Fig. 4.

Brief information source description for a composite plot includes a list of comprising primitive information sources as well as a composite plot caption (see Fig. 2).

4 Quantitative Spectroscopy Ontology Metrics

Users of application data, primarily those located in data collections related to data intensive subject domains, currently meet problems of selection of necessary data, which concern not only the data intension, but also the data quality level. The ontologically described collections are preferable. Such collections can be objectively compared in terms of metrics of the corresponding ontologies. Naturally, the multiplicity of ontology descriptions gives information about a collection of significantly better quality. A certain standard of such a description should arise for each of applied subject domains with time. Below we give an example of the quantitative estimation of the ontology description of resources in the W@DIS.

As a result of the work, a set of spectral data was collected and systematized within the Molecular Spectroscopy IS for several molecules: H_2O, H_2S, $HOCl$, OCS, O_3, SO_2, C_2H_2, CH_4, CO_2, CH_3OH, CO, HBr, HCl, HF, HI, N_2, CH_3Br, CH_3Cl, N_2O, NH_3, NO_2, PH_3, and their isotopologues. The numerical array of spectral data in the

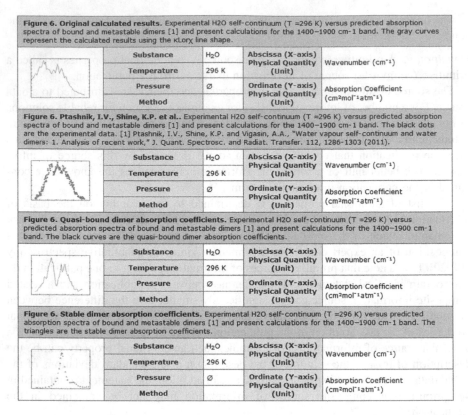

Figure 6. Original calculated results. Experimental H2O self-continuum (T =296 K) versus predicted absorption spectra of bound and metastable dimers [1] and present calculations for the 1400–1900 cm-1 band. The gray curves represent the calculated results using the κLorχ line shape.

	Substance	H$_2$O	Abscissa (X-axis) Physical Quantity (Unit)	Wavenumber (cm⁻¹)
	Temperature	296 K		
	Pressure	Ø	Ordinate (Y-axis) Physical Quantity (Unit)	Absorption Coefficient (cm²mol⁻¹atm⁻¹)
	Method			

Figure 6. Ptashnik, I.V., Shine, K.P. et al.. Experimental H2O self-continuum (T =296 K) versus predicted absorption spectra of bound and metastable dimers [1] and present calculations for the 1400–1900 cm-1 band. The black dots are the experimental data. [1] Ptashnik, I.V., Shine, K.P. and Vigasin, A.A., "Water vapour self-continuum and water dimers: 1. Analysis of recent work," J. Quant. Spectrosc. and Radiat. Transfer. 112, 1286-1303 (2011).

	Substance	H$_2$O	Abscissa (X-axis) Physical Quantity (Unit)	Wavenumber (cm⁻¹)
	Temperature	296 K		
	Pressure	Ø	Ordinate (Y-axis) Physical Quantity (Unit)	Absorption Coefficient (cm²mol⁻¹atm⁻¹)
	Method			

Figure 6. Quasi-bound dimer absorption coefficients. Experimental H2O self-continuum (T =296 K) versus predicted absorption spectra of bound and metastable dimers [1] and present calculations for the 1400–1900 cm-1 band. The black curves are the quasi-bound dimer absorption coefficients.

	Substance	H$_2$O	Abscissa (X-axis) Physical Quantity (Unit)	Wavenumber (cm⁻¹)
	Temperature	296 K		
	Pressure	Ø	Ordinate (Y-axis) Physical Quantity (Unit)	Absorption Coefficient (cm²mol⁻¹atm⁻¹)
	Method			

Figure 6. Stable dimer absorption coefficients. Experimental H2O self-continuum (T =296 K) versus predicted absorption spectra of bound and metastable dimers [1] and present calculations for the 1400–1900 cm-1 band. The triangles are the stable dimer absorption coefficients.

	Substance	H$_2$O	Abscissa (X-axis) Physical Quantity (Unit)	Wavenumber (cm⁻¹)
	Temperature	296 K		
	Pressure	Ø	Ordinate (Y-axis) Physical Quantity (Unit)	Absorption Coefficient (cm²mol⁻¹atm⁻¹)
	Method			

Fig. 4. An example of brief information source description, which corresponds to the composite plot in Fig. 2

Molecular Spectroscopy IS occupies about 80 GB in MySQL database, where most data refer to the H$_2$O molecule and its isotopologues. The MySQL database structure and the size of the numerical data array in the IS are the best choice for real-time meeting the users' information needs. The size of the numerical data array could be reduced by means of additional optimization of the data structure, but then it would have to significantly increase the load on the computing resources of the Molecular Spectroscopy IS. To describe the parts of the complete array, the IS contains about 25 GB of metadata stored in the MySQL database, where the overwhelming majority is the quantitative criteria of data quality derived from the calculations of the values of the correlations between pieces of the numerical data. On the basis of the complete 80-GB data array, ontologies of molecular states and transitions are formed, which are syntax-represented as XML files in RDF/XML notation of the OWL language of about 280 GB in total size. It should be noted that the OWL language has several syntax notations, from the shortest in the Manchester syntax to the longest in the OWL/XML syntax. The relatively verbose RDF/XML syntax was selected for the representation of OWL ontologies in the Molecular Spectroscopy IS because of historical reasons; this choice seemed optimal in the beginning of the work on ontology representations in the

Molecular Spectroscopy IS in 2006. On the basis of the 25-GB array of metadata, a semantic information model is formed as the ontology of information resources, syntax-represented as XML files in the RDF/XML notation of the OWL language of about 3 GB in size. A semantic model of information on spectroscopic plots in the form of the ontology of spectroscopic plots, syntax-represented as an XML file in RDF/XML notation of the OWL language of only 2 MB in size, should be mentioned separately. More complete quantitative information on resources is given in Table 3.

Table 3. Volume of data, metadata, and ontologies in W@DIS IS

List of resources in W@DIS IS	Volume, GB
Data layer	
Spectral data	80.779
Metadata layer	
Metadata	24.772
Ontology layer	
Ontology of information resources on quantitative spectroscopy	3.231
Ontology of molecular states and transitions	280.079
Ontology of scientific graphics on quantitative spectroscopy	0.012
All resources	398.8

The completeness of description of the subject domain and its parts by different applied ontologies is estimated using metrics of the ontologies. Some metrics of the applied ontologies on spectroscopy are given in Table 4.

Table 4. Estimation of the metrics of applied ontologies on quantitative spectroscopy

Ontology	Axiom	Logical axiom	Declaration axioms	Class
OIR	$5.4 * 10^6$	$4.6 * 10^6$	606	324
OMST	$0.97 * 10^9$	$0.9 * 10^9$	68	30
OSG	$1.81 * 10^4$	$1.37 * 10^4$	3690	62
	Object property	Data property	Individual	DL expressivity
OIR	92	355	$1.4 * 10^6$	ALCHON(D)
OMST	13	25	$2.0 * 10^9$	ALC(D)
OSG	17	10	$3.7 * 10^3$	ALCHO(D)

OIR means the ontology of information resources, OMST means the ontology of molecular states and transitions, OSG means the ontology of spectroscopic plots.

5 Conclusion

A brief analysis of the published basic quantitative spectroscopy resources, represented in tabular and graphic forms, is given. The part of these resources, analyzed in this work, includes the published numerical arrays and their properties, resulted from calculations and measurements. These arrays are the solutions of the seven spectroscopy problems. One of these problems deals with measurements and others comprise direct and inverse computational problems. The arrays and their properties combine into the spectral data and information collection in W@DIS. Once the data is uploaded each collection array is described automatically, both in the database and the ontologies.

The reduction problem for the graphic resources ontology is generally solved with a sufficient granularity for the quantitative spectroscopy domain. The concepts for "Data source" and "Information source" are defined and used for describing properties of numerical arrays, represented in tabular and graphic forms. A brief review of the W@DIS ontology metrics is given.

Acknowledgements. The work was financially supported by the Russian Foundation for Basic Research (grant no. 07-13-0411).

References

1. Berners-Lee, T., Hendler, J., Lassila, O.: The semantic web. Sci. Am. **284**(5), 28–37 (2001)
2. Kalinichenko, L., et al.: New Data access challenges for data intensive research in Russia. In: CEUR Workshop Proceedings, vol. 1536, pp. 215–237 (2015)
3. Csaszar, A., Furtenbacher, T., Árendás, P.: Small molecules - big data. J. Phys. Chem. A **120** (45), 8949–8969 (2016). https://doi.org/10.1021/acs.jpca.6b02293
4. Lavrentyev, N.A., Makogon, M.M., Fazliev, A.Z.: Comparison of the HITRAN and GEISA spectral databases taking into account the restriction on publication of spectral data. Atmos. Oceanic Opt. **24**(5), 436–451 (2011)
5. Fazliev, A., Privezentsev, A., Tsarkov, D., Tennyson, J.: Ontology-based content trust support of expert information resources in quantitative spectroscopy. In: Klinov, P., Mouromtsev, D. (eds.) KESW 2013. CCIS, vol. 394, pp. 15–28. Springer, Heidelberg (2013). https://doi.org/10.1007/978-3-642-41360-5_2
6. Voronina, S.S., Privezentsev, A.I., Tsarkov, D.V., Fazliev, A.Z.: An ontological description of states and transitions in quantitative spectroscopy. In: Proceedings of SPIE XX-th International Symposium on Atmospheric and Ocean Optics: Atmospheric Physics, vol. 9292, p. 92920C (2014)
7. Lavrentiev, N.A., Rodimova, O.B., Fazliev, A.Z.: Systematization of graphically plotted published spectral functions of weakly bound water complexes. In: Matvienko, G.G., Romanovski, O.A. (eds.) Proceedings of the SPIE of 22nd International Symposium on Atmospheric and Ocean Optics: Atmospheric Physics, vol. 10035, p. 100350C (2016). https://doi.org/10.1117/12.2249159
8. Tennyson, J., Yurchenko, S.N.: Data structures for ExoMol: molecular line lists for exoplanet and other atmospheres. Mon. Not. R. Astron. Soc. **425**, 21–33 (2012). https://doi.org/10.1111/j.1365-2966.2012.21440.x

9. Tennyson, J., et al.: A database of water transitions from experiment and theory (IUPAC Technical Report). Pure Appl. Chem. **86**(1), 71–83 (2014). https://doi.org/10.1515/pac-2014-5012

10. Keller-Rudek, H., Moortgat, G.K., Sander, R., Sörensen, R.: The MPI-Mainz UV/VIS spectral atlas of gaseous molecules of atmospheric interest. Earth Syst. Sci. Data **5**, 365–373 (2013). https://doi.org/10.5281/zenodo.6951

11. Lavrentiev, N.A., Privezentsev, A.I., Fazliev, A.Z.: Systematization of tabular and graphical resources in quantitative spectroscopy. In: Kalinichenko, L., Manolopoulos, Y., Stupnikov, S., Skvortsov, N., Sukhomlin, V. (eds.) Selected Papers of the XX International Conference on Data Analytics and Management in Data Intensive Domains (DAMDID/RCDL 2018). CEUR Workshop Proceedings, vol. 2277, pp. 25–32 (2018)

12. Akhlyostin, A., et al.: The current status of the W@DIS information system. In: Matvienko, G.G., Romanovski, O.A. (eds.) 22nd International Symposium on Atmospheric and Ocean Optics: Atmospheric Physics. Proceedings of SPIE, vol. 10035, p. 100350D (2016). https://doi.org/10.1117/12.2249235

13. Rodimova, O.B.: Continuum water vapor absorption in the 4000–8000 cm^{-1} region. In: 21st International Symposium Atmospheric and Ocean Optics: Atmospheric Physics. Proceedings of SPIE, vol. 9680, p. 968002 (2015). https://doi.org/10.1117/12.2205332

14. Oparin, D.V., Filippov, N.N., Grigoriev, I.M., Kouzov, A.P.: Effect of stable and metastable dimers on collision-induced rototranslational spectra: carbon dioxide – rare gas mixtures. J. Quant. Spectrosc. Radiat. Transfer **196**, 87–93 (2017). https://doi.org/10.1016/j.jqsrt.2017.04.002

15. Jacquinet-Husson, N., et al.: The 2015 edition of the GEISA spectroscopic database. J. Mol. Spectrosc. **327**, 31–72 (2016). https://doi.org/10.1016/j.jms.2016.06.007

16. Richard, C., et al.: New section of the HITRAN database: collision-induced absorption (CIA). J. Quant. Spectrosc. Radiat. Transfer **113**(11), 1276–1285 (2012). https://doi.org/10.1016/j.jqsrt.2011.11.004

17. Goldenstein, C.S., Miller, V.A., Spearrin, R.M., Strand, C.L.: SpectraPlot.com: integrated spectroscopic modeling of atomic and molecular gases. J. Quant. Spectrosc. Radiat. Transfer **200**, 249–257 (2017). https://doi.org/10.1016/j.jqsrt.2017.06.007

18. Richard, C., Dubernet, M.-L., Moreau, N., Boudon, V.: Spectroscopic databases for the VAMDC portal: new tools and improvements. In: 73rd International Symposium on Molecular Spectroscopy (2018). https://doi.org/10.15278/isms.2018.tf01

19. Kozodoyev, A.V., Kozodoyeva, E.M.: Extensible module "unary operations" in the information system "molecular spectroscopy". Vestnik NSU Ser. Inf. Technol. **13**(1), 46–54 (2015). (in Russian)

20. Kozodoev, A.V., Kozodoeva, E.M.: The binary operations in the information system "molecular spectroscopy". Vestnik NSU Ser. Inf. Technol. **16**(2), 70–77 (2018). https://doi.org/10.25205/1818-7900-2018-16-2-70-77. (in Russian)

Data Models

The Principles and the Conceptual Architecture of the Metagraph Storage System

Valeriy M. Chernenkiy, Yuriy E. Gapanyuk$^{(\boxtimes)}$,
Georgiy I. Revunkov, Ark M. Andreev, Yuriy T. Kaganov,
Ivan V. Dunin, and Maxim A. Lyaskovsky

Informatics and Control Systems Department, Bauman Moscow State Technical
University, Baumanskaya 2-ya, 5, 105005 Moscow, Russia
{chernen,gapyu,revunkov,kaganov.y.t}@bmstu.ru,
arkandreev@gmail.com, johnmoony@yandex.ru,
maksim_lya@mail.ru

Abstract. This paper discusses an approach for active metagraph model storage. The formal definition of the metagraph data model is proposed. The example of data metagraph model is given. The formal definition of the metagraph function and rule agents are discussed. The example of a metagraph rule agent is given. It is shown that the distinguishing feature of the metagraph agent is its homoiconicity which means that it can be a data structure for itself. Thus, the metagraph agent can change both data metagraph fragments and the structure of other metagraph agents. The definition of active metagraph is given. The possible states of active metagraph elements and transitions between them are discussed. The conceptual architecture of the metagraph storage system based on active metagraph is proposed. The approaches for mapping the metagraph model to the flat graph, document-oriented, and relational data models are proposed. The experiments result for storing the metagraph model in different databases are given. It is shown that the flat graph model is most suitable for metagraph storage.

Keywords: Metagraph · Metavertex · Metagraph agent ·
Metagraph function agent · Mctagraph rule agent · Active metagraph ·
Flat graph · Graph database · Document-oriented database · Relational database

1 Introduction

Nowadays models based on complex graphs are increasingly used in various fields of science from mathematics and computer science to biology and sociology. There are currently only graph databases based on flat graph or hypergraph models that are not capable enough of being suitable repositories for complex relations in the domains.

We propose to use a metagraph data model that allows storing more complex relationships than a flat graph model.

The paper is devoted to methods of storage of the metagraph model based on the flat graph, document-oriented, and relational data models. We have tried to offer a general approach to store metagraph data in any database with the mentioned above

© Springer Nature Switzerland AG 2019
Y. Manolopoulos and S. Stupnikov (Eds.): DAMDID/RCDL 2018, CCIS 1003, pp. 73–87, 2019.
https://doi.org/10.1007/978-3-030-23584-0_5

data model. But at the same time, we conducted experiments on several databases. The results of the experiments are presented in the corresponding section.

This paper is an extended version of our paper [1]. Compared to paper [1], sections with new materials "The Active Metagraph and Principles of its Storage" and "The Conceptual Architecture of the Metagraph Storage System" have been added to this article. Section "The Metagraph Agent and its Homoiconicity" is not completely new material, but has been added for reasons of clarity. The discussion about RDF model removed from this version of the paper.

The paper [1] addressed the storage of data metagraphs. This paper offers a holistic view of the metagraph storage, designed to store both metagraph data and metagraph agents. For this, an approach based on active metagraphs is used.

2   The Description of the Metagraph Model

In this section, we will describe the metagraph model. This model may be considered as a "logical" model of the metagraph storage.

A metagraph is a kind of complex network model, proposed by Basu and Blanning [2] and then adapted for information systems description by the present authors [1]. According to [1]:

$$MG = \langle MG^V, MG^{MV}, MG^E \rangle,$$

where MG – metagraph; MG^V – set of metagraph vertices; MG^{MV} – set of metagraph metavertices; MG^E – set of metagraph edges.

A metagraph vertex is described by the set of attributes: $v_i = \{atr_k\}, v_i \in MG^V$, where v_i – metagraph vertex; atr_k – attribute.

A metagraph edge is described by the set of attributes, the source, and destination vertices and edge direction flag:

$$e_i = v \langle s, v_E, eo, \{atr_k\} \rangle, e_i \in MG^E, eo = true|false,$$

where e_i – metagraph edge; v_S – source vertex (metavertex) of the edge; v_E – destination vertex (metavertex) of the edge; eo – edge direction flag ($eo = true$ – directed edge, $eo = false$ – undirected edge); atr_k – attribute.

The metagraph fragment:

$$MG_i = \{ev_j\}, ev_j \in (MG^V \cup MG^{MV} \cup MG^E),$$

where MG_i – metagraph fragment; ev_j – an element that belongs to the union of vertices, metavertices, and edges.

The metagraph metavertex:

$$mv_i = \langle \{atr_k\}, MG_j \rangle, mv_i \in MG^{MV},$$

where mv_i – metagraph metavertex belongs to set of metagraph metavertices MG^{MV}; atr_k – attribute, MG_j – metagraph fragment.

Thus, a metavertex in addition to the attributes includes a fragment of the metagraph. The presence of private attributes and connections for a metavertex is a distinguishing feature of a metagraph. It makes the definition of metagraph holonic – a metavertex may include a number of lower level elements and in turn, may be included in a number of higher-level elements.

From the general system theory point of view, a metavertex is a special case of the manifestation of the emergence principle, which means that a metavertex with its private attributes and connections becomes a whole that cannot be separated into its component parts. The example of metagraph is shown in Fig. 1.

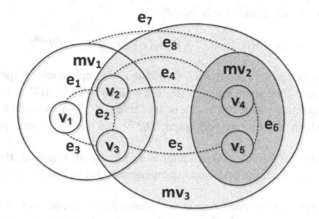

Fig. 1. The example of metagraph

This example contains three metavertices: mv_1, mv_2, and mv_3. Metavertex mv_1 contains vertices v_1, v_2, v_3 and connecting them edges e_1, e_2, e_3. Metavertex mv_2 contains vertices v_4, v_5 and connecting them edge e_6. Edges e_4, e_5 are examples of edges connecting vertices v_2–v_4 and v_3–v_5 respectively and are contained in different metavertices mv_1 and mv_2. Edge e_7 is an example of an edge connecting metavertices mv_1 and mv_2. Edge e_8 is an example of an edge connecting vertex v_2 and metavertex mv_2. Metavertex mv_3 contains metavertex mv_2, vertices v_2, v_3 and edge e_2 from metavertex mv_1 and also edges e_4, e_5, e_8 showing the holonic nature of the metagraph structure. Figure 1 shows that the metagraph model allows describing complex data structures and it is the metavertex that allows implementing emergence principle in data structures.

3   The Metagraph Agent and Its Homoiconicity

The metagraph model is aimed for data and knowledge description. But it is not aimed for data transformation. To solve this issue the metagraph agent (ag^{MG}) aimed for data transformation was proposed in our paper [3].

There are two kinds of metagraph agents: the metagraph function agent (ag^F) and the metagraph rule agent (ag^R). Thus $ag^{MG} = (ag^F | ag^R)$.

The metagraph function agent serves as a function with input and output parameter in the form of metagraph:

$$ag^F = \langle MG_{IN}, MG_{OUT}, AST \rangle,$$

where ag^F – metagraph function agent; MG_{IN} – input parameter metagraph; MG_{OUT} – output parameter metagraph; AST – abstract syntax tree of metagraph function agent in form of metagraph.

The metagraph rule agent is rule-based:

$$ag^R = \langle MG, R, AG^{ST} \rangle, R = \{r_i\}, r_i : MG_j \rightarrow OP^{MG},$$

where ag^R – metagraph rule agent; MG – working metagraph, a metagraph on the basis of which the rules of the agent are performed; R – set of rules r_i; AG^{ST} – start condition (metagraph fragment for start rule check or start rule); MG_j – a metagraph fragment on the basis of which the rule is performed; OP^{MG} – a set of actions performed on metagraph.

The antecedent of the rule is a condition over metagraph fragment, the consequent of the rule is a set of actions performed on metagraph. Rules can be divided into open and closed.

The consequent of the open rule is not permitted to change metagraph fragment occurring in rule antecedent. In this case, the input and output metagraph fragments may be separated. The open rule is similar to the template that generates the output metagraph based on the input metagraph.

The consequent of the closed rule is permitted to change metagraph fragment occurring in rule antecedent. The metagraph fragment changing in rule consequent cause to trigger the antecedents of other rules bound to the same metagraph fragment. But incorrectly designed closed rules system can cause to an infinite loop of metagraph rule agent.

Thus, metagraph rule agent can generate the output metagraph based on the input metagraph (using open rules) or can modify the single metagraph (using closed rules).

The example of metagraph rule agent is shown in Fig. 2. The metagraph rule agent "metagraph rule agent 1" is represented as metagraph metavertex. According to the definition it is bound to the working metagraph MG_1 – a metagraph on the basis of which the rules of the agent are performed. This binding is shown with edge e_4.

The metagraph rule agent description contains inner metavertices corresponds to agent rules (rule 1 ... rule N). Each rule metavertex contains antecedent and consequent inner vertices. In given example mv_2 metavertex bound with antecedent which is shown with edge e_2 and mv_3 metavertex bound with consequent which is shown with edge e_3. Antecedent conditions and consequent actions are defined in the form of attributes bound to antecedent and consequent corresponding vertices.

The start condition is given in the form of attribute "start = true". If the start condition is defined as a start metagraph fragment, then the edge bound start metagraph fragment to agent metavertex (edge e_1 in the given example) is annotated with the attribute "start = true". If the start condition is defined as a start rule, then the rule metavertex is annotated with attribute "start = true" (rule 1 in the given example). Figure 2 shows both cases corresponding to the start metagraph fragment and to the start rule.

The distinguishing feature of the metagraph agent is its homoiconicity which means that it can be a data structure for itself. This is due to the fact that according to definition metagraph agent may be represented as a set of metagraph fragments, and this set can be combined in a single metagraph. Thus, the metagraph agent can change the structure of other metagraph agents.

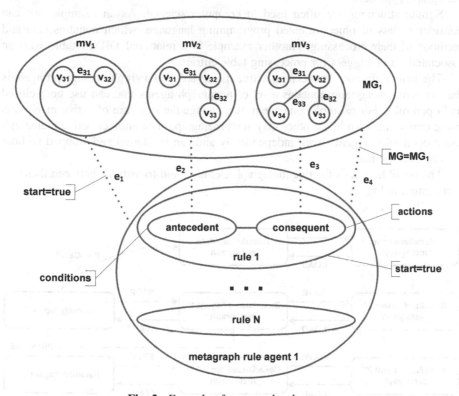

Fig. 2. Example of metagraph rule agent

It should also be noted that efficient pattern matching algorithms used in production systems often use a graph representation of production rules. A well-known example of such an algorithm is RETE. In the case of metagraph approach, the rules of the metagraph agent may be transformed into RETE-network (Alpha and Beta networks) using the higher-level metagraph agent.

4 The Active Metagraph and Principles of Its Storage

In order to combine the data metagraph model and metagraph agent model we propose the concept of "active metagraph":

$$MG^{ACTIVE} = \langle MG^D, AG^{MG} \rangle, AG^{MG} = \{ag_i^{MG}\},$$

where MG^{ACTIVE} – an active metagraph; MG^D – data metagraph; AG^{MG} – set of metagraph agents ag_i^{MG}, attached to the data metagraph.

Thus, active metagraph allows to combine data and processing tools for the metagraph approach.

Similar structures are often used in computer science. As an example, we can consider a class of object-oriented programming language, which contains data and methods of their processing. Another example is a relational DBMS table with an associated set of triggers for processing table entries.

The main difference between an active metagraph and a single metagraph agent is that an active metagraph contains a set of metagraph agents that can use both closed and open rules. For example, one agent may change the structure of active metagraph using closed rules while the other may send metagraph data another active metagraph using open rules. Agents work independently and can be started and stopped without affecting each other.

The possible states of active metagraph elements and transitions between them are represented in Fig. 3.

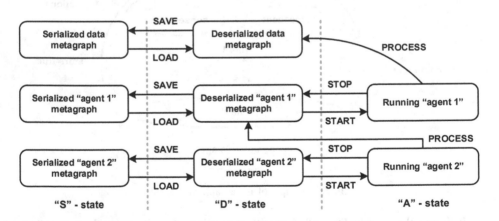

Fig. 3. Possible states of elements of the active metagraph and transitions between them

There are three possible states for active metagraph elements:

1. "S-state" is "serialized" or "stored" state. The serialized active metagraph elements are saved into the store and not ready for processing.
2. "D-state" is "deserialized" state. In this state data metagraph or metagraph agent are ready for processing as passive data structures.

3. "A-state" is "active" state. This state is possible only for metagraph agents. In active state, the metagraph agent may process any metagraph data in "D-state".

It should be noted that "S-state" does not depend on the storage "physical" data model, although different storage models are discussed in the following sections.

Consider the transitions between states. The transitions correspond to the possible operations that can be performed on elements in these states:

- SAVE – serialize metagraph data (which can be data metagraph or agent representation) from the "D-state" to the "S-state".
- LOAD – deserialize metagraph data from the "S-state" to the "D-state" (which is the reverse action for SAVE operation).
- START – starting the execution of metagraph agent on the basis of metagraph agent data representation, i.e. transferring agent from the "D-state" to the "A-state".
- STOP – stopping the metagraph agent (which is the reverse action for START operation).
- PROCESS – any metagraph agent in "A-state" may process any metagraph data in "D-state" (which can be data metagraph or agent representation).

Figure 3 shows an example with data metagraph and two metagraph agents. The "agent 1" changes the structure of data metagraph while "agent 2" changes the structure of "agent 1" using PROCESS operation.

Whereas several active metagraphs may share data metagraph fragments, therefore the PROCESS operation may be used for data-driven communication between agents.

It should also be noted that the active metagraph is a hierarchical structure because the hierarchical structure is the data metagraph included in the definition of the active metagraph.

5 The Conceptual Architecture of the Metagraph Storage System

The conceptual architecture of the metagraph storage system based on active metagraph is represented in Fig. 4.

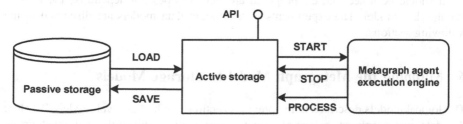

Fig. 4. The conceptual architecture of the metagraph storage system

The passive storage is aimed to store serialized metagraph data that cannot be processed directly. It corresponds to the "S-state" of the active metagraph.

The active storage is aimed to deals with ready for processing metagraph data. It corresponds to the "D-state" of the active metagraph. Let us consider in more detail the features of active storage:

- First of all, active storage is an API for metagraph data processing.
- Active storage makes it possible to process the metagraph data programmatically, which is the basis for the PROCESS operation.
- The metagraph representation of agent stored in active storage may be transformed into an executable format and send to the metagraph agent execution engine (START operation).

Several ways to implement active storage can be suggested:

1. Active storage may be implemented on the basis of the in-memory database. This case has an advantage in the processing speed of metagraph data, but also has a limit on the amount of in-memory data processed. In this case LOAD and SAVE are the physical operations for reading/writing metagraph data from passive storage into the in-memory database.
2. Active storage can be implemented as an add-on over the passive storage without using the intermediate database. In this case, active storage API calls will be translated into the corresponding operations on the passive storage. The LOAD and SAVE operations mean the passive storage reading and writing according to API calls.

The detailed implementation of the active storage is the subject of further research.

The metagraph agent execution engine is aimed to run metagraph agents. It corresponds to the "A-state" of the active metagraph. The running metagraph agent may process any metagraph data in active storage (PROCESS operation). The execution of agent may be stopped, and metagraph representation of agent is saved to the active storage (STOP operation).

The proposed conceptual architecture deals with the homoiconic nature of metagraph agents. Agents may be processed as metagraph data structures and executed as programs. Thus, both data metagraph and metagraph agents may be stored in active and passive storages.

It should be noted that the proposed architecture does not depend on the passive storage data model. The experiments with different data models are discussed in the following sections.

6   Mapping the Metagraph Model to Storage Models

The logical models described in the previous sections are higher-level models. To store the data metagraph or metagraph agent representation efficiently, we must create mappings from "logical" model to "physical" models used in different databases.

In this section, we will consider the metagraph model mappings to the flat graph model, document model, and relational model.

6.1   Mapping Metagraph Model to the Flat Graph Model

The main idea of this mapping is to flatten the hierarchical metagraph model.

Of course, it is impossible to turn a hierarchical graph model into a flat one directly. The key idea to do this is to use multipartite graphs [4].

Consider there is a flat graph:

$$FG = \langle FG^V, FG^E \rangle,$$

where FG^V – set of graph vertices; FG^E – set of graph edges.

Then a flat graph FG may be unambiguously transformed into bipartite graph BFG:

$$BFG = \langle BFG^{VERT}, BFG^{EDGE} \rangle,$$

$$BFG^{VERT} = \langle FG^{BV}, FG^{BE} \rangle,$$

$$FG^V \leftrightarrow FG^{BV}, FG^E \leftrightarrow FG^{BE},$$

where BFG^{VERT} – set of graph vertices; BFG^{EDGE} – set of graph edges. The set BFG^{VERT} can be divided into two disjoint and independent sets FG^{BV} and FG^{BE} and there are two isomorphisms $FG^V \leftrightarrow FG^{BV}$ and $FG^E \leftrightarrow FG^{BE}$. Thus, we transform the edges of graph FG into a subset of vertices of graph BFG. The set BFG^{EDGE} stores the information about relations between vertices and edges in graph FG.

It is important to note that from bipartite graph point of view there is no difference whether original graph FG oriented or not, because edges of the graph FG are represented as vertices and, orientation sign became the property of the new vertex.

From the general system theory point of view, transforming edge into vertex, we consider the relation between entities as a special kind of higher-order entity that includes lower-level vertices entities.

Now we will apply this approach of flattening to metagraphs. In the case of metagraph we use not bipartite but tripartite target graph TFG:

$$TFG = \langle TFG^{VERT}, TFG^{EDGE} \rangle,$$

$$TFG^{VERT} = \langle TFG^V, TFG^E, TFG^{MV} \rangle,$$

$$TFG^V \leftrightarrow MG^V, TFG^E \leftrightarrow MG^E, TFG^{MV} \leftrightarrow MG^{MV}.$$

The set TFG^{VERT} can be divided into three disjoint and independent sets TFG^V, TFG^E, TFG^{MV}. There are three isomorphisms between metagraph vertices, metavertices, edges and corresponding subsets of TFG^{VERT}: $TFG^V \leftrightarrow MG^V$, $TFG^E \leftrightarrow MG^E$, $TFG^{MV} \leftrightarrow MG^{MV}$. The set TFG^{EDGE} stores the information about relations between vertices, metavertices, edges in original metagraph.

Consider the example of flattening metagraph model represented in Fig. 5. The vertices, metavertices, and edges of original metagraph are represented with vertices of different shapes.

From the general system theory point of view, emergent metagraph elements such as vertices, metavertices, edges are transformed into independent vertices of the flat graph.

The proposed mapping may be used for storing metagraph data in graph or hybrid databases such as Neo4j or ArangoDB.

It is important to note that flattening metagraph model does not solve all problems for graph database usage. Consider the example of a query using the Neo4j database query language "Cypher":

```
(n1:Label1)-[rel:TYPE] → (n2:Label2)
```

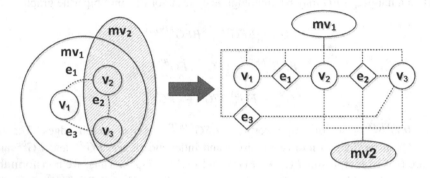

Fig. 5. The example of metagraph for flattening (shown on the left) and the example of flattened metagraph (shown on the right)

One can see that used notation is RDF-like and suppose that graph edges are named. But flatten metagraph model does not use named edges because metagraph edges are transformed into vertices.

Thus, query languages of flat graph databases are not suitable for the metagraph model because they blur the semantics of the metagraph model.

6.2 Mapping Metagraph Model to the Document Model

From the general system theory point of view, emergent metagraph elements such as vertices, metavertices, edges should be represented as independent entities.

In the previous subsection, we use flat graph vertices for such a representation. But instead of graph vertices, we can also represent independent entities as documents for the document-oriented database. Flat graph edges are represented as relations via id-idrefs between documents.

For the sake of clarity, we use the Prolog-like predicate textual representation. This representation may be easily converted into JSON or XML formats because it is compliant with JSON semantics and contains nested key-value pairs and collections.

The classical Prolog uses the following form of the predicate: $predicate(atom_1, atom_2, \cdots, atom_N)$. We used an extended form of predicate where along with atoms predicate can also include key-value pairs and nested predicates: $predicate(atom, \cdots, key = value, \cdots, predicate(\cdots), \cdots)$. The mapping of metagraph model fragments into predicate representation is described in details in [3].

The proposed textual representation may be used for storing metagraph data in a document-oriented database or text or document fields of the relational database using JSON or XML formats.

6.3 Mapping Metagraph Model to the Relational Model

Nowadays NoSQL databases are very popular. But traditional relational databases are still the most mature solution and widely used in information systems. Therefore, we also need the relational representation of the metagraph model. There are two ways to store metagraphs in a relational database.

The first way is to use a pure relational schema. In this case, the proposed metagraph model may be directly or with some optimization transformed into the database schema. The tables vertices, metavertices, edges may be used. Figure 6 contains a graphical representation of such a schema using a PostgreSQL database. The table "metavertex" contains the representation of vertices and metavertices. The table "relation" contains the representation of edges.

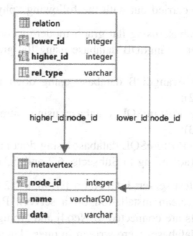

Fig. 6. The database schema for pure relational metagraph representation

The second way is to use document-oriented possibilities of a relational database. For example, the latest versions of PostgreSQL database provide such a possibility. The Fig. 7 contains a graphical representation of such a schema.

In this case, vertices, metavertices, and edges are stored as XML or JSON documents in relational tables. The drawback of this approach is id-idrefs storage between documents.

Figure 7 shows an example where id-idrefs are stored inside a binary JSON "data" field. In the case of a relational database, we have to parse data fields and process id-idrefs links programmatically which decrease the overall system performance.

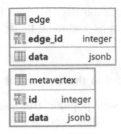

Fig. 7. The database schema for document-relational metagraph representation

7 The Experiments

In this section, we present experiments results for storing the metagraph "logical" model in several databases with different "physical" data models.

It should be noted that these are just entry-level experiments that should help to choose the right data model prototype for the metagraph storage system.

The experiments were carried out with the following "physical" data models:

- Neo4j – the Neo4j database using flat graph model (according to Subsect. 6.1);
- ArangoDB(graph) – the ArangoDB database using flat graph model (according to Subsect. 6.1);
- ArangoDB(doc) – the ArangoDB database using document-oriented model (according to Subsect. 6.2);
- PostgreSQL(rel) – the PostgreSQL database using pure relational schema (according to Subsect. 6.3);
- PostgreSQL(doc) – the PostgreSQL database using document-oriented possibilities of relational database (according to Subsects. 6.2 and 6.3).

The characteristics of test server: Intel Xeon E7-4830 2.2 GHz, 4 GB RAM, 1 Tb SSD, OS Ubuntu 16.04 (clean installation on a server). Python 3.5 was used for running test scripts. Scripts are connected to Neo4j and ArangoDB via official Python drivers. Queries to these databases were written in query languages (Cypher and AQL respectively) without ORM and executed by Python drivers. However, queries for PostgreSQL were made with SQLAlchemy ORM in order to simplify database manipulations from the python script. In all cases, the database was generated by scripts in CSV-format. The database was reloaded from the dump after every test, which modified the state of the database.

Each operation was repeated several times to get the average time of execution.

The experimental dataset consisted of 1 000 000 vertices, randomly connected with 1 000 000 edges. Each vertex of the dataset included one random integer attribute and one random string attribute of fixed length. For read operations (selecting hierarchy), additional ten vertices of fixed structure (100 nested levels) were added to the dataset to get an average time of 10 reads.

The results of tests are represented in the Table 1. This is the test time in milliseconds; the less value is better. The best results are marked in bold. If the best result is approximately the same for several databases, then all these cases are marked in Table 1.

Let's make intermediate conclusions on the basis of the considered results of experiments.

It is necessary to recognize the Neo4j implementation as inefficient compared to other cases. But this is not a disadvantage of the flat graph model itself, because the graph implementation in ArangoDB is quite efficient.

The inserting, updating and deleting operations are very efficient in PostgreSQL (both relational and document-oriented schemas) and ArangoDB (document-oriented schema), but this is not the case for hierarchical selecting which is typical metagraph operation.

The time for hierarchical selecting for graph databases (both Neo4j and ArangoDB) is comparable to the time of other tests while the time for hierarchical selecting for relational and document-oriented databases is several times longer than the time of other tests.

Thus, if the system architect is forced to use the metagraph passive storage based on a relational or document-oriented database, then hierarchical selecting queries should be the subject of careful optimization.

Summarizing, we can say that, provided an effective graph database is used, the flat graph model is most suitable for metagraph storage.

Table 1. The results of tests (test time in milliseconds, the less value is better).

Test case	Neo4j	ArangoDB (graph)	ArangoDB (doc)	PostgreSQL (rel)	PostgreSQL (doc)
Inserting vertex to the existing metavertex	40	**2**	5	8	6
Inserting vertex to the metagraph	253	**3**	**3**	**3**	4
Inserting edge to the metagraph	148	32	**7**	8	**6**
Updating existing vertex value	267	**5**	**5**	**3**	9
Deleting vertex from the existing metavertex	45	**6**	**5**	**6**	9
Deleting edge from the existing metavertex	57	**6**	16	9	**6**
Selecting hierarchy of 100 related metavertices	45	**5**	323	218	187
The number of best results	**0/7**	**6/7**	**4/7**	**4/7**	**3/7**

8 Related Work

Nowadays, there is a tendency to complicate the graph database data model. An example of this tendency is the HypergraphDB [5] database. As the name implies, HypergraphDB uses the hypergraph as a data model. The reasoning capabilities are implemented via integration with TuProlog.

Another interesting project is a GRAKN.AI [6] aimed for AI purpose that explicitly combines graph-based and ontology-based approach for data analysis. The flat graphs and hypergraphs may be used as a data model. The Graql query language is used both for data manipulation and reasoning.

It was shown in our paper [7] that the metagraph is a holonic graph model whereas the hypergraph is a near flat graph model that does not fully implement the emergence principle. Therefore, the hypergraph model doesn't fit well for complex data structures description.

But in fact, HypergraphDB and GRAKN.AI use a hierarchical hypergraph model which is suitable for complex networks description.

Nowadays the semantic web approach for knowledge representation is widely used. In this case, the Resource Description Framework (RDF) is used as the data model, and SPARQL is used as the query language. RDFS (RDF Schema) and OWL (OWL2) are used as ontology definition languages, built on the base of RDF. Using RDFS and OWL, it is possible to express various relationships between ontology elements (class, subclass, equivalent class, etc.) [8]. For RDF persisting and SPARQL processing, special storage systems are used, e.g., Apache Jena.

But unfortunately, the RDF approach has several limitations for complex situation description which are considered in details in [1]. The root of limitations is the absence of the emergence principle in the flat graph RDF model. The metagraph model addresses RDF limitations in a natural way without emergence loss. Therefore, despite the prevalence of the RDF model, we consider the development of a storage system for the metagraph model as an important task.

9 Conclusions

The models based on complex graphs are increasingly used in various fields of science from mathematics and computer science to biology and sociology. Nowadays, there is a tendency to complicate the graph database data model in order to support the complexity of the domains.

We propose to use a metagraph data model that allows storing more complex relationships than a hypergraph data model and RDF model.

The metagraph agents are aimed for metagraph data transformation. The distinguishing feature of the metagraph agent is its homoiconicity. The metagraph agent can change the structure of other metagraph agents.

The active metagraph is aimed to combine the data metagraph model and the metagraph agent model.

The proposed conceptual architecture deals with the homoiconic nature of metagraph agents. Agents may be processed as metagraph data structures and executed as programs. Thus, both data metagraph and metagraph agents may be stored in active and passive storages.

The metagraph model may be mapped to the flat graph model, the document model and the relational model. The main idea of this mapping it the flattening metagraph to the flat multipartite graph. Then the flat graph may be represented as a document model or relational model.

The experiments results show that flat graph model is most suitable for metagraph storage.

In the future, it is planned to develop a metagraph data manipulation language and design a stable version of the metagraph storage based on a flat graph database.

References

1. Chernenkiy, V., Gapanyuk, Y., Kaganov, Y., Dunin, I., Lyaskovsky, M., Larionov, V.: Storing metagraph model in relational, document-oriented, and graph databases. In: Proceedings of the XX International Conference on Data Analytics and Management in Data Intensive Domains (DAMDID/RCDL 2018), Moscow, Russia, 9–12 October 2018, pp. 82–89 (2018). http://ceur-ws.org/Vol-2277/paper17.pdf
2. Basu, A., Blanning, R.: Metagraphs and Their Applications. Integrated Series in Information Systems, vol. 15. Springer, New York (2007). https://doi.org/10.1007/978-0-387-37234-1
3. Chernenkiy, V.M., Gapanyuk, Y.E., Nardid, A.N., Gushcha, A.V., Fedorenko, Y.S.: The hybrid multidimensional-ontological data model based on metagraph approach. In: Petrenko, A.K., Voronkov, A. (eds.) PSI 2017. LNCS, vol. 10742, pp. 72–87. Springer, Cham (2018). https://doi.org/10.1007/978-3-319-74313-4_6
4. Chartrand, G., Zhang, P.: Chromatic Graph Theory. Discrete Mathematics and its Applications. Chapman & Hall/CRC, Boca Raton (2009)
5. HyperGraphDB website. http://hypergraphdb.org/
6. GRAKN.AI website. https://grakn.ai/
7. Chernenkiy, V., Gapanyuk, Y., Revunkov, G., Kaganov, Y., Fedorenko, Y., Minakova, S.: Using metagraph approach for complex domains description. In: Proceedings of the XIX International Conference on Data Analytics and Management in Data Intensive Domains (DAMDID/RCDL 2017), Moscow, Russia, 9–13 October 2017, pp. 342–349 (2017). http://ceur-ws.org/Vol-2022/paper52.pdf
8. Allemang, D., Hendler, J.A.: Semantic Web for the Working Ontologist. Effective Modeling in RDFS and OWL, 2nd edn. Morgan Kaufmann/Elsevier, Burlington/Amsterdam (2011)

Data Analysis in Astronomy

Data Analysis of Astronomy

Evaluation of Binary Star Formation Models Using Well-Observed Visual Binaries

Oleg Malkov[1]([📧]) [ID], Dmitry Chulkov[1] [ID], Yikdem Gebrehiwot[3,4] [ID],
Dana Kovaleva[1] [ID], Nikolay A. Skvortsov[2] [ID], Alexey Sytov[1],
Solomon Belay Tessema[5], Alexander Tutukov[1],
and Lev Yungelson[1] [ID]

[1] Institute of Astronomy of Russian Academy of Sciences, Moscow, Russia
{malkov, chulkov, dana, sytov, atutukov, lry}@inasan.ru
[2] Institute of Informatics Problems, Federal Research Center "Computer Science
and Control", Russian Academy of Sciences, Moscow, Russia
nskv@mail.ru
[3] Entoto Observatory and Research Center, Addis Ababa, Ethiopia
yikdema16@gmail.com
[4] College of Natural and Computational Sciences, Mekelle University,
Mekelle, Ethiopia
[5] Entoto Observatory and Research Center Astronomy and Astrophysics,
Ethiopian Space Science and Technology Institute, Addis Ababa, Ethiopia
tessemabelay@gmail.com

Abstract. Creation of the Galaxy model describing formation and evolution of binary stars requires generating and testing hypotheses related to the process of formation of binary stars and distributions of their characteristics. A set of hypotheses can be generated on the basis of a number of publications that suggested the formation of binary systems. We describe the project aimed at finding initial distributions of binary stars over masses of components, mass ratios of them, semi-major axes and eccentricities of orbit, and also pairing scenarios by means of Monte-Carlo modeling of the sample of visual binaries of luminosity class V with a set of additional restrictions, so it can be considered as free of observational incompleteness effects. We present results which allow rejecting some estimated initial distributions of visual binary star parameters.

Keywords: Binary stars · Stellar formation · Modeling

1 Introduction

Creating models that simulate research objects determines possibilities to obtain new knowledge about these objects. Research is not limited to the analysis of observed characteristics of physical systems but allows to evaluate physical characteristics, which can not be observed by scientific instruments today, but which could cause definite values of the observed characteristics. Majority of stars accessible for detailed observational study appear to be binary ones. Interaction between binary star components in the course of their evolution results in a rich variety of astrophysical

© Springer Nature Switzerland AG 2019
Y. Manolopoulos and S. Stupnikov (Eds.): DAMDID/RCDL 2018, CCIS 1003, pp. 91–107, 2019.
https://doi.org/10.1007/978-3-030-23584-0_6

phenomena and objects. Study of the structure and evolution of binary stars is one of the most actively developing fields of modern astrophysics.

Single star formation and evolution have been modeled and studied. Theoretical modeling allows obtaining models of stars clusters with statistical properties agreeing with observations. Known investigations were devoted to star formation in a narrow range of spectral classes of stars, certain evolutionary status stars. But there is no model taking into account formation of interacting star systems in a part of the Milky Way galaxy.

To model a population of binary stars, we study the birth function (BF) which defines a statistical dependence of the number of stars born on a set of parameters. Most important, BF is, first, a benchmark for the theories of star formation and, second, the base for the estimates of the number of objects in the models of different stellar populations and model rates of various events, e.g., supernovae explosions.

Hypotheses related to the process of binary star formation and distributions of their initial characteristics should be generated and testing. A set of hypotheses can be derived from publications that have estimated some aspects of binary star formation. Verification of the hypotheses will allow revealing constraints of the parameters of the star birth function of binary systems and building a consistent model of the history of star formation of binary systems in the Galaxy. Hypothesis testing is performed by comparing the model with observational data.

We consider the problem of determining the fundamental astrophysical parameters of the binary star formation history as a solution of an inverse problem, based on distributions of binary stars on the observed characteristics. As an inverse problem, it is ill-posed and can be solved only under the condition of imposing additional restrictions on desired distributions.

The aim of the paper is presenting results of the assessment of BF by means of comparison of results of Monte-Carlo model of the local population of field visual binaries with their observed sample.

We probe, for a given type of stars, whether the synthetic dataset differs significantly due to the change of initial fundamental distributions, and how the change of every distribution affects it. For this purpose, we compare synthetic populations for different pairing functions and particular sets of fundamental functions. We attempt to find whether certain initial distributions or combinations of initial distributions result in synthetic datasets incompatible with observational data at certain significance level and, on the contrary, whether certain initial distributions or combinations of initial distributions provide synthetic dataset best compatible with observational data, hopefully, at certain significance level.

The paper is organized as follows. The model of initial fundamental distributions of binary star parameters is described in Sect. 2. Some considerations on the choice of theoretical models are discussed in Sect. 3. Observational data sources and limitations used to compare with modeled data are chosen in Sect. 4. Results and conclusions are presented in Sect. 5. In Sect. 6 we outline the plans of future studies.

The paper restructures and extends the report presented in [25]. The introduction (Sect. 1) has been rewritten to emphasize the importance of modeling as a research approach when investigated object characteristics cannot be observed directly, and to describe briefly the binary star birth function as a research object and a way of investigating it. Some modeling details are replaced from the introduction to the model description section (Sect. 2). Section 2 has been extended with a detailed description of the birth function constituents for visual binary stars and with a description of an approach to generating random values with a specified distribution. In Sect. 6 future plans of research were refined, and some specificities of other observational types of binary stars are described. Finally, a conclusion was added.

2 The Model

To make a model of binary star formation history, it is necessary to determine and investigate constituents (parameters) of the star formation function. The following estimations are proposed in various publications as fundamental parameters for binary star formation:

- estimating scenarios of binary star mass formation and selection of respective parameter pairing [19];
- possible type of distributions of binary systems and their components by masses;
- distribution of binary systems by semi-major axes of orbits [30];
- possible type of distribution of binary systems over the eccentricities of orbits [37].

It should be investigated which parameter distributions are fundamental for binary systems – masses of components, a ratio of their masses, total mass of a system, etc. It is still an open problem. There is an estimate that formation of wide and close binaries occur in different scenarios. Various combinations of these hypotheses were included in considerations as fundamental ones.

Distributions of mass ratios of binary star components ($q = M_2/M_1$) can vary from decreasing from zero to one to increasing in the same range or even bimodal ones for different types of binaries. Mass ratio distributions differ for close and wide binary systems too. The distribution of primary and secondary components themselves for wide binary systems seems to be the same as the initial mass function of single stars. Depending on chosen scenario of wide pairs, mass ratio is defined by the masses of the components or represented by a distribution in the form of a power function.

Preferable model of the distribution of binaries by semi-major axes of orbits is a linear distribution in the logarithmic scale, however, independence of this distribution from component masses is not proved. A lognormal distribution for this parameter is possible too due to dynamic changes of orbits and/or selection effects. There is an assumption that unimodal initial distribution on semi-major axes of orbits cannot correspond to observed ones. For wide pair formation, we have tested different parameters of the linear distribution in logarithmic scale.

Distribution of wide binary systems by orbit eccentricities is stable. Close binary stars with orbit periods less than ten days are circularized soon. So eccentricity distribution can be an important parameter describing initial distributions. For wide pairs, we show that the initial distribution of eccentricities weakly affects the resulting distribution. Eccentricities o close pairs are determined much more massively and can serve as an observational parameter, its distribution needs to be analyzed later.

Different kinds of distributions can be used to generate initial values of binary star parameters (such as semi-major axes, orbit eccentricities, star formation rate, and others). There are distribution laws modeled as constant, linear, power and exponential functions. The main principle of generating random values with a given distribution law is as follows. Let $f(x)$ is a probability density function, x_{min} and x_{max} are limit values of the argument, k is a normalization factor such that

$$\int_{x_{min}}^{x_{max}} kf(x)dx = 1$$

The cumulative probability is determined as

$$F(x) = \int kf(x)dx = \Phi(x) + C$$

where $\Phi(x)$ is the antiderivative of $f(x)$, and C is a constant computed from the initial conditions $F(x_{min}) = 0$.

$R = F(x)$ is randomly generated in the range from 0 to 1. A target value of x is calculated as an inversion of $\Phi(x)$ for randomly generated R:

$$x(F) = \Phi^{-1}(R - C)$$

thus its differential distribution (i.e., the probability density of x) is determined by $f(x)$.

The following form of the birth function (BF) for visual binaries was suggested by Vereshchagin et al. [38]:

$$d^3N \propto M_1^{-2.5}dM_1 \cdot d\log a \cdot q^{-2.5}dq\left(year^{-1}\right) \tag{1}$$

where M_1 is mass of the primary component expressed in solar units ($M_\odot$), $q = M_2/M_1$ is mass ratio of components, a is semi-major axis of a component orbit expressed in solar units. In the present study, we consider different combinations of fundamental functions [29] describing the distribution of stars and generate random values of binary star parameters with respective distributions. As a "minor" characteristics, we consider eccentricity of orbits e.

To simulate stellar pairs we use different pairing functions (scenarios), mostly taken from Kouwenhoven's list [19]. It includes random pairing and other scenarios, where two of the four parameters (primary mass, secondary mass, total mass of the system, mass ratio) are randomized, and others are calculated. Table 1 contains a short summary of the used pairing functions.

Masses of the of components or total masses of the binaries were drawn randomly from Salpeter [35] or Kroupa [24] initial mass functions (IMF), separation a was drawn from one of the following distributions: $\propto a^{-1}$, $\propto a^{-1.5}$, and $\propto a^{-2}$, and eccentricity e was distributed assuming the following options: (i) all orbits are circular, (ii) eccentricities obey thermal distribution $f_e(e) = 2e$, and (iii) equiprobable distribution $f_e(e) = 1$. We adopt random orbit orientation. Mass ratio q, when needed, is randomly drawn from $\propto q^\beta$ distribution, where β is adopted to be -0.5, 0 or 0.5. The lower limit for q is determined by mass limits $[0.08 \cdots 100]$ $M_\odot$. Certain pairing functions, such as RP, PCRP, PSCP and TPP, do not allow independent random distribution of mass ratios, it is calculated from masses of components.

Table 1. Summary of considered pairing functions (scenarios).

Abbreviation	Full name	Scheme
RP	Random Pairing	rand($M_1, M_2, [M_{min} \cdots M_{max}]$); sort($M_1, M_2$); calc($q$)
PCRP	Primary Constrained Random Pairing	rand($M_1, [M_{min} \cdots M_{max}]$); rand($M_2, [M_{min} \cdots M_{max}], M_1 = const$ until $M_2 < M_1$); calc(q)
PCP	Primary Constrained Pairing	rand(M_1, q); calc(M_2)
SCP	Split-Core Pairing	rand($M_{tot}, [2M_{min} \cdots 2M_{max}]$); rand($q$); calc($M_1, M_2$)
PSCP	Primary Split-Core Pairing	rand($M_{tot}, [2M_{min} \cdots 2M_{max}]$); rand($M_1, [0.5(M_1 + M_2) \cdots M_{max}]$, until $M_1 < M_{tot}$); calc(M_2); calc(q)
TPP	Total Primary Pairing	rand($M_{tot}, [2M_{min} \cdots 2M_{max}]$); rand($M_1, [M_{min} \cdots M_{max}]$, until $M_1 < M_{tot}$); calc(M_2); sort(M_1, M_2); calc(q)

Note: M_{tot}, M_1, M_2 are total mass of the binary, primary mass and secondary mass, respectively; M_{min}, M_{max} are lower ($0.08 M_\odot$) and upper ($100 M_\odot$) limits set for masses; $q = M_2/M_1$ is mass ratio. The meaning of abbreviations is the following: "rand" – randomizing, "calc" – calculation, "sort" – sorting.

Table 2 contains a short summary of initial distributions used in the modeling. Some cells are empty because the pre-planned distributions are not implemented as yet. The total number of possible combinations of initial distributions considered as yet is 144, equal to the number of possible combinations of s, m, q, a, e, in Table 2 and regarding that $s0$ and $s5$ scenarios do not imply independent initial distribution over q (see Table 1). Any combination of distributions listed in Table 2 can be conveniently referred, for instance, as "s2m0q5a1e0".

Table 2. Summary of applied initial distributions

sN	Scenario (s)	mN	IMF (m)	qN	Mass ratio (q)	aN	Semi-major axis (a)	eN	Eccentricity (e)
0	RP	0	Salpeter	0	Flat, $f = 1$	0	Power, $f \sim a^{-1}$	0	Thermal, $f = 2e$
		1	Kroupa			1	Power, $f \sim a^{-1.5}$	1	Delta, $f = \delta(0)$
2	PCP					2	Power, $f \sim a^{-2}$	2	Flat, $f = 1$
3	SCP								
				4	Power, $f \sim q^{-0.5}$				
5	TPP			5	Power, $f \sim q^{0.5}$				

The model also accounts for star formation rate, stellar evolution and takes into account observational selection effects. Besides, our model allows obtaining estimates for the fraction of binary stars that remain unseen for different reasons and is observed as single objects and to investigate how these fractions depend on the initial distributions of parameters. Such estimates are important, for instance, as an approach toward recovering actual multiplicity fraction, mass hidden in binaries, as well as toward models of different stellar populations.

To account for star formation rate we adopt $SFR(t) = 15e^{-t/7}$, where the time t is expressed in Gyr (Yu and Jeffery [39]). Disc age is assumed to be equal to 14 Gyr.

Currently, we consider the following stellar evolutionary stages: MS-star, red giant, white dwarf, neutron star, black hole. The objects in the two latter stages do not produce visual binaries (though they contribute to the statistics of pairs, observed as single stars, see Sect. 5.2 below). We do not consider brown dwarfs and pre-MS stars here, as they are extremely rarely observed among visual binaries and their multiplicity rate is substantially lower than for more massive stars [1]. As we deal with wide pairs only, we assume the components to evolve independently. To calculate the evolution of stars and their observational properties we used analytical expressions derived by Hurley et al. [18] and assumed solar metallicity for all generated stars.

To normalize the number of simulated objects, we use estimates of stellar density in the solar neighborhood, based on recent Gaia results [6].

For the generated objects, we determine observational parameters, in particular, the brightness of components, their evolutionary stage and projected separation. Then we apply a filter to select a sample of stars, which can be compared with observational data (see the next section).

3   Some Reflections Concerning Selection of Models

In the selection of trial initial distributions for the model, we adopted the following approach: we started with well established or widely used in the literature functions for $f(M), f(a), f(e)$ and then stepped aside from them to test, whether the algorithm would be able to feel a difference at all. We preferred simple analytical expressions, supposing we would pass to more complicated ones later if we find it necessary.

Thus, we use traditional Salpeter's IMF [35] along with the much more recent and generally accepted Kroupa's one [24]. In spite of the statement by Duchêne and Kraus [9] that no observed dataset agrees with random pairing scenario, we use the latter among other ones.

On the other hand, for semi-major axis distribution, we applied as yet only commonly used power law parametrization, with the particular case of a log-log flat distribution known as "Öpik's law" [30]. Validity of $f_a \propto a^{-1}$ law up to $a \approx 4600$ AU, which is close to a_{max} of our refined sample of visual binaries, was confirmed by Popova et al. [31] and Vereshchagin et al. [38] who analyzed the data in the amended 7th Catalog of Spectroscopic Binaries [22] and IDS, respectively. Poveda et al. [32], found that Öpik's distribution matches with a high degree of confidence binaries with $a \lesssim 3500$ AU (but we note, that selection effects which hamper discovery of the widest systems were not considered, contrary to the abovementioned studies). We also stress, after Heacox [16], that Gaussian distribution of separations encountered in the literature (e.g., Duquennoy and Mayor [8], Raghavan et al. [33]) is an artifact of data representation. Like Poveda et al. [32], we reject Gaussian distribution of stellar separations, since it is hard to envision currently a star-formation process leading to such a distribution.

As for the eccentricity distribution, from a physical point of view, one usually prefers in theoretical simulations the "thermal" law $f(e) \sim 2e$ [2], though in observational datasets one finds, e.g., that the eccentricity distribution of wide binaries contains more orbits with $e < 0.2$ and less orbits with $e > 0.8$ (Tokovinin and Kiyaeva [37]) or a flat distribution in the $e = [0.0 \cdots 0.6]$ range and declining one for larger e [33].

Having in mind the difficulties hampering determination of eccentricities from observations and numerous selection effects, we probe three quite different model distributions: "thermal", flat, and single-valued with $e = 0$ for all stars.

The very selection of fundamental parameters for initial distribution is arguable. For instance, primary and secondary masses were considered as fundamental parameters for MS binaries by Malkov [27] and pre-MS binaries by Malkov and Zinnecker [26], while Goodwin [13] has argued that system mass is the more fundamental physical parameter to use. We do not reject the possibility to choose and investigate other parameters as fundamental ones in the course of further work.

4   Observational Data for Comparison

To compare our simulations with observational data, we use the most comprehensive list of visual binaries WCT [20], compiled on the base of the largest original catalogs WDS [28], CCDM [7] and TDSC [10]. These data were refined or corrected for mistaken data, optical pairs, effects of higher degrees of multiplicity, sorted by luminosity class (primarily, to select pairs with both components on the main-sequence), and appended by parallaxes. A refined dataset for comparison was selected from the data, so as to avoid regions of observational incompleteness in the space of observational parameters. The procedure of dataset compilation and analysis described in details in [20, 21] was improved due to use of new trigonometric parallaxes from TGAS DR1 Gaia [11] that allowed to re-obtain constraints to avoid regions of observational incompleteness.

Visual binaries are observed, mostly, in the immediate solar vicinity. Therefore, we consider them to be distributed up to the distance of 500 pc in the radial direction and according to a barometric function along z. The scale height z for the stars of different spectral types and, respectively, masses was studied, e.g., in [5, 12, 14, 23, 34]. Synthesizing results of these studies, we assume $|z| = 340$ pc for low-mass ($\leq 1 M_\odot$) stars, 50 pc for high-mass ($\geq 1 M_\odot$) stars, and linear $|z| - \log M$ relation for intermediate masses. For such a small volume we can neglect the radial gradient [17]. We also ignore interstellar extinction.

Out of simulated objects we select pairs, satisfying the same observational constraints, as the refined observational set does, namely: projected separation $2 < \rho < 200$ arcsec, primary component visual magnitude $V_1 < 9.5^m$, secondary component visual magnitude $V_2 < 11.5^m$, magnitude difference $\Delta V \equiv |V_2 - V_1| \leq 4^m$ (henceforth, "synthetic dataset"). For the purposes of correct comparison, we also limit the refined set of observational data by $500 \sim$ pc distance. The data for A0V-K4V stars presented by Bovy [6] give 0.01033 stars per pc^3. This means that in the 500 pc sphere we generate about 43300 pairs of stars.

We construct distributions of synthetic datasets over the following parameters: primary and secondary magnitude, magnitude difference, projected separation, parallax. Then we compare the synthetic distributions with refined observational ones using χ^2 two-sample test. We deem, the better result of comparison, the closer our assumptions on pairing scenarios, initial distributions of masses, mass ratio, separation and eccentricity are to reality. The refined set of observational data contains $N = 1089$ stars. To compare them properly with results of our simulations we need to use histograms with $n = 5 \log N$ bins [36], i.e., 15 ones.

5 Results and Discussion

5.1 Star Formation Function

Comparison of our simulations with observational data allows us to make the following conclusions on initial distributions. Even before application of statistical tests, we should meet a strong and evidently important criterion of validity of the tested combination of initial distributions, namely the agreement between the number of binaries in the simulated datasets and the observed number of visual pairs. This number depends on initial distributions of fundamental variables and changes between 0 and about 15000; an exception is a distribution over e which affects the volume of simulated datasets only mildly. Thus, if our accepted normalization [6], along with other used assumptions regarding spatial distribution of visual binaries in solar vicinity is valid, we can exclude certain combinations of initial distributions, based purely on the number of binaries in synthetic dataset. However, our present observational dataset volume (1089 pairs) is limited to binaries having MK spectral classification. Thus, we are careful and do not rely exclusively on this criterion because we allow certain freedom due to simplifications and possible incomplete account of selection effects while constructing the refined observational dataset, as well as to vagueness of theoretical notions on solar vicinity population. This is why we do take into account both number and two sample χ^2 criteria. Nevertheless, one can definitely reject those combinations of initial distributions that lead to the number of binary stars in a synthetic observational dataset significantly less than 1000 (taking present dataset volume $N_{obs} - \sqrt{N_{obs}} \approx 1056$ as lower limit).

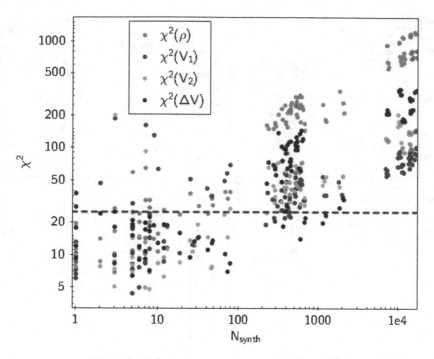

Fig. 1. Distribution of resulting χ^2 statistics over a number of pairs in the synthetic dataset. Every set of initial distributions of the 144 processed ones results in 4 dots of different colour in this plot. The dashed line marks 5% confidence level of the null-hypothesis (the dots over it correspond to the sets of initial distributions that are rejected at the level of 95%, based on the used observational sample).

Figure 1 represents how the resulting χ^2 statistics are distributed versus a number of pairs in the synthetic datasets. The results do not allow us to select "the best" initial distributions over every parameter, but rather to prefer some combinations of initial distributions to others. One may see that no combination leading to an acceptable number of pairs in the synthetic dataset would give acceptable distribution over angular distances between components, while magnitude difference and, in some cases, distribution over primary magnitudes, are reproduced better for the same initial conditions. Below there are some figures providing examples of how the same distribution over certain parameter, in different combinations with other initial distributions, leads to better or worse agreement with the observational dataset.

Figure 2 represents an example of how different combinations of initial distributions change the resulting synthetic datasets and their agreement with observational one. Four figures demonstrate, in turn, which values of N_{synth}, χ^2 correspond to different initial scenarios ($s0$, $s2$, $s4$, $s5$, see Tables 1 and 2), IMFs ($m0$, $m1$), mass ratio initial distribution ($q0$, $q4$, $q5$, applicable solely for the $s2$, $s3$ scenarios), and distribution over semi-major axes $a0$, $a1$, $a2$. Scenarios $s0$ and $s5$ do not involve independent distribution over q; it is generated as an outcome of the pairing function and IMF, this is why the q-panel contains less dots than the other ones.

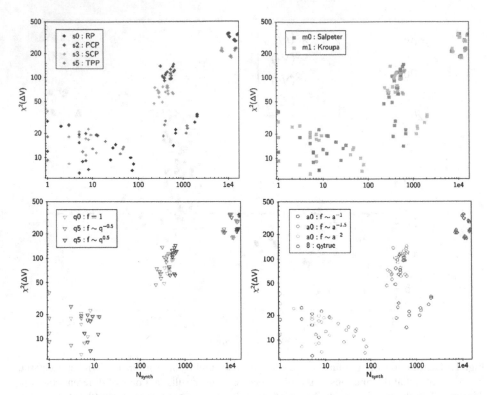

Fig. 2. Distribution of resulting χ^2 statistics for magnitude difference ΔV vs. a number of pairs in the synthetic datasets, depending on various initial distributions, from top to bottom: pairing scenarios (see Tables 1 and 2), IMFs, distributions over mass ratio (applicable solely for scenarios $s2$, $s3$), and semi-major axes.

Figure 3 shows how the distribution over observational parameter magnitude difference changes with the change of one initial distribution (pairing scenarios, IMF, distribution over semi-major axes). The distribution over ΔV for the observational dataset serves as a benchmark.

Based on combination of the two (number and statistical) criteria, we may state the following.

For the considered observational dataset, RP and TPP pairing scenarios, $s0$ and $s5$ (see Tables 1 and 2), respectively, produce a group of results that seems acceptable in respect of the number of "observed" binaries in the synthetic dataset and, simultaneously, leads to acceptable χ^2 values at least for two observable distributions (V_1 and ΔV).

None of the probed combinations of initial distributions can reproduce observational distribution over angular distance between components adequately (see Fig. 1). The cause may lay either with selection effects, that still remain unaccounted for (and then the reconsideration of observational sample is necessary) or in need of other initial distributions.

Kroupa and Salpeter IMF's lead to different number of pairs in the synthetic dataset, however, neither this difference nor χ^2 statistics allows definite choice between

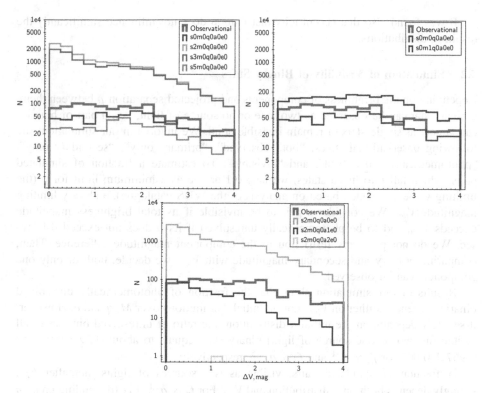

Fig. 3. Distributions of resulting synthetic datasets over magnitude difference ΔV for the combinations of initial distributions differing only in pairing scenarios (see Tables 1 and 2), IMFs, and semi-major axes. The distribution over ΔV for the observational dataset plotted by the bold red line serves as a benchmark. (Color figure online)

them. Kroupa IMF looks slightly more promising, than Salpeter one, however, more accurate conclusion should be postponed, as these two IMF differ actually only in the low-mass region, and the majority of visual binaries in our observational dataset presumably have masses around 1 to 3 $M_\odot$. The comparison in a low-mass region is needed here.

Also, we cannot make a definite conclusion on the mass ratio q distribution. The q-distributions that we have analyzed in the present study show significant difference only in the low-q region (below $q < 0.5$). In the compilative sample of visual binaries used to construct our benchmark dataset, however, binaries with large magnitude differences (and, thus, low q) are severely underrepresented. This is why we limit refined observational sample so that pairs with low q are excluded. For this reason, we cannot come to a definite conclusions concerning selection of q-distribution based on this observational sample.

As to the semi-major axes (a) distribution, we have found that power law functions steeper than $a^{-1.5}$ can be excluded from further consideration. Figures 2 and 3 demonstrate that initial distribution $a2$ ($f \sim a^{-2}$, Table 2) leads to inappropriately low volume of the synthetic dataset.

It was found also that eccentricity distribution does not influence significantly the resulting distributions.

5.2   Simulation of Visibility of Binary Stars

Depending on the brightness of components and projected separation ρ between them, a binary star can be observed as two, one or no source of light, i.e., a part of binaries can appear as single stars or remain invisible at all. We involve in our simulations the following observational states: "both observed", "primary only", "secondary only", "photometrically unresolved", and "invisible". To estimate a fraction of simulated pairs, which fall into listed states, we take 0.1 arcsec as a minimum limit for ρ (the limiting value is selected based on analysis of the WDS catalogue), and vary limiting magnitude V_{lim}. We consider a pair to be invisible if its total brightness magnitude exceeds V_{lim}, and to be photometrically unresolved if its ρ does not exceed 0.1 arcsec. We do not pose any restriction to the component magnitude difference. Then, comparing primary and secondary magnitude with V_{lim}, we decide, both or only one component can be observed.

Results of our simulation show that the fraction of photometrically unresolved binaries depends neither on V_{lim}, nor on initial distributions over M, q and e. However, it severely depends on the initial a-distribution: the ratio of unresolved binaries to all visible (as two or one source of light) binary stars equals to about 0.59 ± 0.01 and 0.967 ± 0.003 for $f_a \propto a^{-1}$ and $f_a \propto a^{-1.5}$, respectively.

A fraction of simulated pairs, visible as two sources of lights, hereafter F_{PS}, strongly depends both on a-distribution and V_{lim}. For $f_a \propto a^{-1}$, F_{PS} (depending on q, m and e distributions) varies from 0.01 to 0.19 for $V_{lim} = 16^m$ and from 0.04 to 0.26 for $V_{lim} = 20^m$. F_{PS} values are about ten times lower for $f_a \propto a^{-1.5}$.

Finally, the fraction of simulated stars observed as a single source of light depends on the F_{PS} as follows: $0.4 - F_{PS}$ for $f_a \propto a^{-1}$ and $0.03 - 0.7 \times F_{PS}$ for $f_a \propto a^{-1.5}$, with no significant dependence on other parameters.

We should note that simple compatibility of synthetic data from initial distributions with observational data is not ultimate evidence of adequate modeling since the observational data are far from being comprehensive. However, the initial distributions we use in our simulations are obtained, checked and used by many other authors, and this suggests that our conclusions are fairly reliable.

6   Future Plans

We presented here results proving that we need a more thorough investigation of the models and comparison with observations to accomplish the task.

To make confident conclusions on BF of binary stars, we need to make a comparison of our simulations with other sets of observational data for wide binaries, to cover wider regions of stellar parameters.

One of the important further steps is to extend our study to close binary systems. We will involve basic ideas on evolution of interacting binaries in our simulations, and, consequently, will take into consideration other types of binaries for comparison.

In particular close binary systems of the most representative observational classes eclipsing and spectroscopic binary stars - will be studied. For components of these systems, sufficient sets of parameters can be determined from observations to be used to study the process of star formation.

Choosing the preferable combinations of initial scenarios for close binary systems at different stages of evolution and comparing them with the results obtained for wide pairs we will search for a solution to the inverse problem of restoring the parameters of the star-formation function for the entire range of parameters of binary systems. It will allow us to make conclusions on the universality or difference in the parameters of the star formation function for wide and close binaries, on BF of more massive stars (as, simulating visual binaries, we deal mostly with moderate-mass stars), to consider more distant objects, and to involve final stages of stellar evolution into consideration. Having a number of Monte-Carlo simulations representing various observational datasets, we should be able to check if the approximate formula (1) needs reconsideration or remains valid.

Different samples of binary stars of different observational and evolutionary types (wide and close pairs) will be used as observational data to work with binary stars not observable photometrically as binary ones and with close binary stars. Analysis of selection effects for each sample will result in restricting an area in the parameters free of the observational incompleteness. It is also planned to use the second data release of the Gaia space mission and containing high-precision parallax, which will allow estimating the absolute values of the linear parameters of the systems quite accurately and massively. We aim to consider other parameters as fundamental for initial distributions, e.g., total mass of a binary, angular momentum of a pair, and so on.

Besides the χ^2 two sample test, we plan to consider other statistical methods (e.g., Kolmogorov-Smirnov two sample test) for more reliable interpretation of comparison of our simulation results with observations.

6.1   Modeling Different Observational Types of Binaries

We will analyze a sample of orbital binaries. Orbital binary is a visual binary with known orbital elements and known distance. Orbital binaries are essential objects for determining dynamical and physical properties of stars, especially masses, through a combined analysis of photometric and astrometric data. Along with double-lined eclipsing binaries, orbital binaries with known distances are the only types of detached binary systems that enable one to determine stellar masses and semi-major axes of orbits. Orbital binaries are quite numerous: current (2019) version of the principal catalog of orbital binaries, ORB6 [15] contains data on about 3000 orbits. The catalog represents the best collections of published data on orbital binaries, however, it contains neither spectral classification nor parallax information, so the necessary data should be added from other catalogs or from literature. To compile a "refined" subset of the orbit list for further modeling, at least the following selection criteria should be used: (i) the existence of trigonometric parallax, (ii) the absence of third companion, and (iii) the high quality of orbit. For these systems, stellar masses can be calculated independently from dynamical, photometric and spectral data. Comparison of these values can show a discrepancy and, consequently, can indicate a false input parameter.

After necessary corrections distributions of orbital binaries along observational parameters can be constructed and compared with modeled ones (after taking into account principal selection effects). The resulting distributions will represent a complete sample in a "semi-major axis - primary brightness - magnitude difference" parameter space, and these distributions can be used to construct the initial mass function and star formation history of wide binaries.

Another promising observational type of binaries for comparison of a model with observations is eclipsing binaries. They are also very numerous, the Catalogue of eclipsing variables, CEV [4], contains information on some 7000 classified binaries. Eclipsing binaries provide the methods by which fundamental stellar parameters (such as mass, radius, luminosity, etc.) can be independently estimated. It is especially true for cases when one observes binary star as eclipsing and spectroscopic (when lines of both components are seen in the composite spectra) simultaneously. In this case, the precision of the derived values will be as high as several percentages. However, the number of such systems is only about 2% of full amount of known eclipsing binaries and will not increase considerably in the future. On the other hand, there is a huge amount of eclipsing binaries discovered during ground-based and space surveys, which still remain unstudied. We will use available information on new binaries for the determination of their evolutionary stage and for searching the unusual systems which belong to the rare evolution stages. For these purposes, we have developed the method for the assessment of evolutionary status of eclipsing binaries using light-curve parameters and spectral classification [3]. The procedure was applied to the list of eclipsing binaries, which were collected in CEV. About 4000 of binaries were classified successfully but also it was found that some systems cannot be classified at all or multi-valued classification is possible. To find out the reason we have checked all these binaries with the literature and stated the three problems: (i) obsolete or unconfirmed values of observational parameters can obstruct the classification; (ii) contradictory values of parameters can lead to uncertain classification; and (iii) extreme and unusual systems sometimes cannot be classified. These problems can be solved with new additional observations and/or investigation. After that, a cataloged sample can be compared with the modeled one.

7 Conclusion

We have described results of modeling initial distributions of binary stars over masses of components, mass ratios, semi-major axes and eccentricities of orbit, and also pairing scenarios by means of Monte-Carlo method. Models have been compared with the sample of about 1000 visual binaries of luminosity class V with Gaia DR1 TGAS trigonometric parallax larger than 2 mas, limited by $2 \leq \rho \leq 200$ arcsec, $V_1 \leq 9.5^m$, $V_2 \leq 11.5^m$, $\Delta V \leq 4^m$, which can be considered as free of observational incompleteness effects. Modeling allowed to reject some known estimates of parameter distributions of binary stars. Future plans of research including other observational types of binary stars have been defined.

Acknowledgments. We are grateful to T. Kouwenhoven, A. Malancheva and D. Trushin for helpful discussions and suggestions. The work was partially supported by the Program of fundamental researches of the Presidium of RAS (P-28), the Russian Foundation for Basic Research (grants 18-07-01434, 18-29-22096, 19-07-01198).

This research has made use of the VizieR catalog access tool and the SIMBAD database operated at CDS, Strasbourg, France, the Washington Double Star Catalog maintained at the U.S. Naval Observatory, NASA's Astrophysics Data System Bibliographic Services, Joint Supercomputer Center of the Russian Academy of Sciences, and data from the European Space Agency (ESA) mission Gaia (https://www.cosmos.esa.int/gaia), processed by the Gaia Data Processing and Analysis Consortium (DPAC, https://www.cosmos.esa.int/web/gaia/dpac/consortium).

References

1. Allers, K.N.: Brown dwarf binaries. In: Richards, M.T., Hubeny, I. (eds.) International Astronomical Union. From Interacting Binaries to Exoplanets: Essential Modeling Tools, vol. 7, no. 282, pp. 105–110 (2012). https://doi.org/10.1017/s1743921311027086
2. Ambartsumian, V.: On the statistics of double stars (Trans. by D.W. Goldmith). Astron. Zh. **14**, 207 (1937)
3. Avvakumova, E.A., Malkov, O.Yu.: Assessment of evolutionary status of eclipsing binaries using light-curve parameters and spectral classification. Mon. Not. R. Astron. Soc. **444**(2), 1982–1992 (2014). https://doi.org/10.1093/mnras/stu1572
4. Avvakumova, E.A., Malkov, O.Yu., Kniazev, A.Y.: Eclipsing variables: catalogue and classification. Astron. Notes **334**, 860–865 (2013). https://doi.org/10.1002/asna.201311942
5. Bahcall, J.N., Soneira, R.M.: The universe at faint magnitudes. I. Models for the galaxy and the predicted star counts. Astrophys. J. Suppl. Ser. **44**, 73–110 (1980). https://doi.org/10.1086/190685
6. Bovy, J.: Stellar inventory of the solar neighbourhood using Gaia DR1. Mon. Not. R. Astron. Soc. **470**(2), 1360–1387 (2017). https://doi.org/10.1093/mnras/stx1277
7. Dommanget, J., Nys, O.: CCDM (Catalog of Components of Double and Multiple stars), VizieR On-line Data Catalog: I/274 (2002)
8. Duquennoy, A., Mayor, M.: Multiplicity among solar-type stars in the solar neighbourhood. II - Distribution of the orbital elements in an unbiased sample. Astron. Astrophys. **248**(2), 485–524 (1991)
9. Duchêne, G., Kraus, A.: Stellar multiplicity. Ann. Rev. Astron. Astrophys. **51**, 269 (2013)
10. Fabricius, C., Høg, E., Makarov, V.V., Mason, B.D., Wycoff, G.L., Urban, S.E.: The Tycho double star catalogue. Astron. Astrophys. **384**(1), 180–189 (2002). https://doi.org/10.1051/0004-6361:20011822
11. Gaia Collaboration, Prusti, T., de Bruijne, J.H.J., et al.: The Gaia mission. Astron. Astrophys. **595**, A1:1–A1:36 (2016). https://doi.org/10.1051/0004-6361/201629272
12. Gilmore, G., Reid, N.: New light on faint stars – III. Galactic structure towards the South Pole and the Galactic thick disc. Mon. Not. R. Astron. Soc. **202**(4), 1025–1047 (1983)
13. Goodwin, S.P.: Binary mass ratios: system mass not primary mass. Mon. Not. R. Astron. Soc. Lett. **430**(1), 6–9 (2013). https://doi.org/10.1093/mnrasl/sls037
14. Gould, A., Bahcall, J.N., Flynn, Ch.: Disk M dwarf luminosity function from HST star counts. Astrophys. J. **465**, 759 (1996)

15. Hartkopf, W.I., Mason, B.D., Worley, C.E.: The 2001 US naval observatory double star CD-ROM. II. The fifth catalog of orbits of visual binary stars. Astron. J. **122**(6), 3472–3479 (2001)
16. Heacox, W.D.: Of logarithms, binary orbits, and self-replicating distributions. Publ. Astron. Soc. Pac. **108**, 591–593 (1996)
17. Huang, Y., Liu, X., Zhang, H., Yuan, H., Xiang, M., Chen, B., et al.: On the metallicity gradients of the Galactic disk as revealed by LSS-GAC red clump stars. Res. Astron. Astrophys. **15**(8), 1240 (2015)
18. Hurley, J.R., Pols, O.R., Tout, C.A.: Comprehensive analytic formulae for stellar evolution as a function of mass and metallicity. Mon. Not. R. Astron. Soc. **315**(3), 543–569 (2000)
19. Kouwenhoven, M.B.N., Brown, A.G.A., Goodwin, S.P., Portegies Zwart, S.F., Kaper, L.: Pairing mechanisms for binary stars. Astron. Nachr.: Astron. Notes **329**(9–10), 984–987 (2008)
20. Kovaleva, D.A., Malkov, O.Yu., Yungelson, L.R., Chulkov, D.A., Yikdem, G.M.: Statistical analysis of the comprehensive list of visual binaries. Baltic Astron. **24**, 367–378 (2015)
21. Kovaleva, D.A., Malkov, O.Yu., Yungelson, L.R., Chulkov, D.A.: Visual binary stars: data to investigate the formation of binaries. Baltic Astron. **25**, 419–426 (2016)
22. Kraitcheva, Z., Popova, E., Tutukov, A., Yungelson, L., Kraitcheva, Z., et al.: Catalogue of physical parameters of spectroscopic binary stars. Bull. Inf. du Centre de Donnees Stellaires **19**, 71 (1980)
23. Kroupa, P.: The distribution of low-mass stars in the disc of the Galaxy. Cambridge University, UK (1992)
24. Kroupa, P.: On the variation of the initial mass function. Mon. Not. R. Astron. Soc. **322**(2), 231–246 (2001)
25. Malkov, O.Yu., et al.: Insight into binary star formation via modelling visual binaries datasets. In: Kalinichenko, L., Manolopoulos, Y., Stupnikov, S., Skvortsov, N., Sukhomlin, V. (eds.) Selected Papers of the XX International Conference on Data Analytics and Management in Data Intensive Domains (DAMDID/RCDL 2018), vol. 2277, pp. 98–106. CEUR (2018). http://ceur-ws.org/Vol-2277/paper19.pdf
26. Malkov, O.Yu., Piskunov, A.E., Zinnecker, H.: On the luminosity ratio of pre-main sequence binaries. Astron. Astrophys. **338**, 452–454 (1998)
27. Malkov, O.Yu., Zinnecker, H.: Binary stars and the fundamental initial mass function. Mon. Not. R. Astron. Soc. **321**(1), 149–154 (2001)
28. Mason, B.D., Wycoff, G.L., Hartkopf, W.I., Douglass, G.G., Worley, C.E.: The Washington Visual Double Star Catalog. VizieR Online Data Catalog: B/wds (2014)
29. Massevich, A., Tutukov, A.: Stellar eEvolution: Theory and Observations. Nauka, Moscow (1988). (in Russian)
30. Öpik, E.: Statistical studies of double stars: on the distribution of relative luminosities and distances of double stars in the harvard revised photometry north of declination −31 deg. In: Publications of the Tartu Astrofizica Observatory, vol. 25 (1924)
31. Popova, E.I., Tutukov, A.V., Yungelson, L.R.: Study of physical properties of spectroscopic binary stars. Astrophys. Space Sci. **88**(1), 55–80 (1982)
32. Poveda, A., Allen, C., Hernández-Alcántara, A.: Halo wide binaries and moving clusters as probes of the dynamical and merger history of our Galaxy. In: Hartkopf, B., Guinan, E., Harmanec, P. (eds.) Binary Stars as Critical Tools and Tests in Contemporary Astrophysics, IAU Symposium, no. 240, p. 417. Cambridge University Press, Cambridge (2007)
33. Raghavan, D., McAlister, H.A., Henry, T.J., et al.: A survey of stellar families: multiplicity of solar-type stars. Astrophys. J. Suppl. Ser. **190**, 1 (2010)
34. Reed, B.C.: New estimates of the scale height and surface density of OB stars in the solar neighborhood. Astron. J. **120**(1), 314 (2000)

35. Salpeter, E.E.: The luminosity function and stellar evolution. Astrophys. J. **121**, 161–167 (1955)
36. Shtorm, R.: Probability Theory. Mathematical Statistics. Statistical Quality Control. 368 p. Mir, Moscow (1970). (in Russian)
37. Tokovinin, A., Kiyaeva, O.: Eccentricity distribution of wide binaries. Mon. Not. R. Astron. Soc. **456**(2), 2070–2079 (2016)
38. Vereshchagin, S., Tutukov, A., Yungelson, L., Kraicheva, Z., Popova, E.: Statistical study of visual binaries. Astrophys. Space Sci. **142**(1–2), 245–254 (1988). Colloquium on Wide Components in Double and Multiple Stars
39. Yu, S., Jeffery, C.S.: The gravitational wave signal from diverse populations of double white dwarf binaries in the Galaxy. Astron. Astrophys. **521**, A85 (2010)

Text Search and Processing

Proximity Full-Text Search by Means of Additional Indexes with Multi-component Keys: In Pursuit of Optimal Performance

Alexander B. Veretennikov(✉) ⓘD

Ural Federal University, Yekaterinburg, Russia
alexander@veretennikov.ru

Abstract. Full-text search engines are important tools for information retrieval. In a proximity full-text search, a document is relevant if it contains query terms near each other, especially if the query terms are frequently occurring words. For each word in a text, we use additional indexes to store information about nearby words that are at distances from the given word of less than or equal to the *MaxDistance* parameter. We showed that additional indexes with three-component keys can be used to improve the average query execution time by up to 94.7 times if the queries consist of high-frequency occurring words. In this paper, we present a new search algorithm with even more performance gains. We consider several strategies for selecting multi-component key indexes for a specific query and compare these strategies with the optimal strategy. We also present the results of search experiments, which show that three-component key indexes enable much faster searches in comparison with two-component key indexes.

Keywords: Full-text search · Search engines · Inverted indexes ·
Additional indexes · Proximity search · Term proximity ·
Information retrieval

1 Introduction

A search query consists of several words. The search result is a list of documents containing these words. In [1], we discussed a methodology for high-performance proximity full-text searches and a search algorithm. With the application of additional indexes [1], we improved the average query processing time by a factor of 94.7 when queries consist of high-frequency occurring words.

In this paper, we present the following new results.

We present a new search algorithm in which we can improve the performance even more than it was improved in [1].

We present the results of search experiments that prove that three-component key indexes can be used to improve the average query execution time by up to 15.6 times in comparison with two-component key indexes when queries consist of high-frequency occurring words.

© Springer Nature Switzerland AG 2019
Y. Manolopoulos and S. Stupnikov (Eds.): DAMDID/RCDL 2018, CCIS 1003, pp. 111–130, 2019.
https://doi.org/10.1007/978-3-030-23584-0_7

In modern full-text search approaches, it is important for a document to contain search query words near each other in order to be relevant to the context of the query, especially if the query contains frequently occurring words. The impact of the term-proximity is integrated into modern information retrieval models [2–5].

Words appear in texts at different frequencies. The typical word frequency distribution is described by Zipf's law [6]. An example of words' occurrence distribution is shown in Fig. 1. The horizontal axis represents different words in decreasing order of their occurrence in texts. On the vertical axis, we plot the number of occurrences of each word.

The full-text search task can be solved with inverted indexes [7–9]. With ordinary inverted indexes, for each word in the indexed document, we store in the index the record (ID, P), where ID is the identifier of the document and P is the position of the word in the document. Let P be an ordinal number of the word in the document.

For proximity full-text searches, we need to store the (ID, P) record for all occurrences of any word in the indexed document. These (ID, P) records are called "postings".

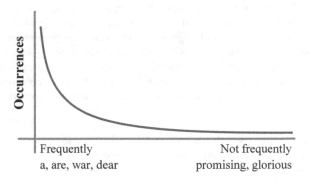

Fig. 1. Example of a word frequency distribution.

Therefore, the query search time is proportional to the number of occurrences of the queried words in the indexed documents. Consequently, to evaluate a search query that contains high-frequency occurring words, a search system needs much more time (see Fig. 1, on the left side) than a query that contains ordinary words (see Fig. 1, on the right side).

A full-text query is a "simple inquiry", and accordingly [10], to prevent the interruption of the thought continuity of the user, the query results must be produced within two seconds. In this context, we present the following problem. It is common to have a full-text search engine that can usually evaluate a query within 1 s. of time. However, it works very slowly, for example, requiring 10–30 s, for a query that contains frequently occurring words.

We can illustrate this problem by the following example. We downloaded pgdvd042010.iso from the Project Gutenberg web page, which contains their files as of April 2010, and we indexed its content using Apache Lucene 7.4.0 and Apache Tika

1.18. We indexed approximately 64 thousand documents with a total length of approximately 13 milliard characters (a relatively small number). We indexed all words. Then, we evaluated the following queries using the equipment from Sect. 4.1 of the current paper:

"Prince Hamlet" ~ 4 – this search took 172 ms, and

"to be or not to be" ~ 4 – this search took 21 s.

The suffix "~ 4" instructs Lucene to search such texts in which the queried words contain no more than 4 other words between them.

To improve the search performance, early-termination approaches can be applied [11, 12]. However, early-termination methods are not effective in the case of proximity full-text searches [1]. It is difficult to combine the early-termination approach with the integration of term-proximity information into relevance models.

Another approach is to create additional indexes. In [13, 14], the authors introduced some additional indexes to improve the search performance, but they only improved phrase searches.

With our additional indexes, an arbitrary query can be evaluated very quickly.

In this paper, we present a new and more effective approach that extends the method from [15]. In the new approach, we try to select the optimal configuration of multi-component key indexes for a specific query. The major extension is shown in the "3.3 Index selection" section, and the results of new experiments are presented.

2 Lemmatization and Lemma Type

2.1 Word Type

In [16], we defined three types of words.

Stop Words: Examples include "and", "at", "or", "not", "yes", "who", "to", "war", "time", "man" and "be". In a stop-word approach, these words are excluded from consideration, but we do not do so. In our approach, we include information about all words in the indexes.

We cannot exclude a word from the search because a high-frequency occurring word can have a specific meaning in the context of a specific query [1, 14]; therefore, excluding some words from consideration can induce search quality degradation or unpredictable effects [14].

Let us consider the query example "who are you who". The Who are an English rock band, and "Who are You" is one of their songs. Therefore, the word "Who" has a specific meaning in the context of this query.

Frequently Used Words: These words are frequently encountered but convey meaning. These words always need to be included in the index. Examples include "beautiful", "red", and "hair".

Ordinary Words: This category contains all other words. Examples include "glorious" and "promising".

2.2 Lemmatization

We employ a morphological analyzer for lemmatization. For each word in the dictionary, the analyzer provides a list of numbers of lemmas (i.e., basic or canonical forms). For a word that does not exist in the dictionary, its lemma is the same as the word itself. Some words have several lemmas. For example, the word "mine" has two lemmas, namely, "mine" and "my".

We use a combined Russian/English dictionary with approximately 200 thousand Russian lemmas and 92 thousand English lemmas.

We define three types of lemmas: stop lemmas, frequently used lemmas and ordinary lemmas. We sort all lemmas in decreasing order of their occurrence frequency in the texts. We call this sorted list the *FL*-list. The number of a lemma in the *FL*-list is called its *FL*-number. Let the *FL*-number of a lemma w be denoted by $FL(w)$.

The first *SWCount* most frequently occurring lemmas are stop lemmas. The second *FUCount* most frequently occurring lemmas are frequently used lemmas. All other lemmas are ordinary lemmas. *SWCount* and *FUCount* are the parameters. We use *SWCount* = 700 and *FUCount* = 2100 in the experiments presented.

If an ordinary lemma, $q,$ occurs in the text so rarely that $FL(q)$ is irrelevant, then we can say that $FL(q) = \sim$. We denote by "$\sim$" some large number.

2.3 Index Type

We create indexes of different types for different types of lemmas. Let *MaxDistance* be a parameter that can take a value of 5, 7 or even greater.

The expanded (f, s, t) index or three-component key index [1, 17] is the list of occurrences of the lemma f for which lemmas s and t both occur in the text at distances that are less than or equal to the *MaxDistance* from f.

We create an expanded (f, s, t) index only for the case in which $f \leq s \leq t$. Here, f, s, and t are all stop lemmas. Each posting includes the distance between f and s in the text and the distance between f and t in the text.

The expanded (w, v) index or two-component key index [18–20] is the list of occurrences of the lemma w for which lemma v occurs in the text at a distance that is less than or equal to the *MaxDistance* from w.

The lemma types considered are as follows: for w, frequently used, and for v, frequently used or ordinary. Each posting includes the distance between w and v in the text.

Other types of additional indexes are described in [1].

3 A New Search Algorithm

3.1 The Search Algorithm General Structure

Our search algorithm is described in Fig. 2. Let us consider the search query "who are you who". After lemmatization, we have the following query:

[who] [are, be] [you] [who]. The word "are" has two lemmas in our dictionary. With *FL*-numbers: [who: 293] [are: 268, be: 21] [you: 47] [who: 293].

To use three-component key indexes, this query must be divided into two subqueries [1]:

Q1: [who: 293] [are: 268] [you: 47] [who: 293], and
Q2: [who: 293] [be: 21] [you: 47] [who: 293].

We can say that lemma "who" > "you" because *FL*(who) = 293, *FL*(you) = 47, and 293 > 47. We use the *FL*-numbers to establish the order of the lemmas in the set of all lemmas.

In [1], we defined several query types depending on the types of lemmas that they contain and the different search algorithms for these query types. The query does not need to be divided into a set of subqueries for all query types.

In this paper, we consider subqueries that consist only of stop lemmas.

After step 2, we evaluate the subqueries in the loop.

After all subqueries are evaluated, their results need to be combined into the final result set.

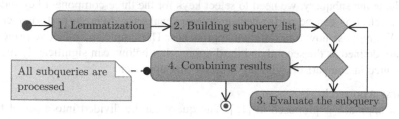

Fig. 2. UML diagram of the search algorithm general structure.

3.2 Subquery Evaluation

The algorithm for the subquery evaluation, when the subquery consists of only stop lemmas, is described in Fig. 3.

We need to select the three-component key indexes required to evaluate the subquery. For all selected indexes, we need to create an iterator object. The iterator object for the key (*f*, *s*, *t*) is used to read the posting list of the (*f*, *s*, *t*) key from the start to the end. The iterator object, *IT*, has the method *IT.Next*, which reads the next record from the posting list.

The iterator object, *IT*, has the property *IT.Value*, which contains the current record (*ID*, *P*, *D1*, *D2*). Consequently, *IT.Value.ID* is the *ID* of the document containing the key, and *IT.Value.P* is the position of the key in the document.

For the two postings of *A* = (*A.ID*, *A.P*, *A.D1*, *A.D2*) and *B* = (*B.ID*, *B.P*, *B.D1*, *B.D2*), we define that *A* < *B* when one of the following conditions is met: *A.ID* < *B.ID* or (*A.ID* = *B.ID* and *A.P* < *B.P*). The records (*ID*, *P*, *D1*, *D2*) are stored in the posting list for the given key in increasing order.

The goal of the *Equalize* procedure is to ensure that all iterators have an equal value of *Value.ID* = *DID*. Afterwards, we can perform the search in the document with identifier *Value.ID*. The *Equalize* procedure is described in [1].

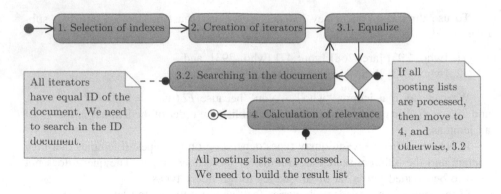

Fig. 3. UML diagram of a subquery evaluation.

3.3   Index Selection

To evaluate the subquery, we need to select keys for the three-component key indexes. The key selection can be performed in different ways for different performance outcomes. We propose now four different approaches. The simple and effective principles, which are defined as the second and third approaches below, can significantly increase performance in comparison with the original first approach.

The First Approach

The first approach is proposed in [15]. The query can be divided into a set of three-component keys. Let the first three lemmas of the query define the first key. Let the next three lemmas of the query define the second key, and so on.

For the cases when the length of the query is not an exact multiple of 3, the last key is always defined by the last three lemmas of the query.

All selected keys must be normalized.

For example, let us consider the subquery [who] [are] [you] [who]. We can use the keys (who, are, you) and (are*, you*, who).

For any three stop lemmas, f, s and t, we have the (f, s, t) index only for the case in which $f \leq s \leq t$. We call the (f, s, t) key with the aforementioned condition the normalized key. The normalized keys here are (you, are, who) and (you*, are*, who).

Let us consider the search query "Who are you and why did you say what you did" and its subquery [who] [are] [you] [and] [why] [do] [you] [say] [what] [you] [do].

In fact, we can find this query in Cecil Forester Scott's novel "Lord Hornblower".

We can use the (who, are, you), (and, why, do), (you, say, what), and (what*, you, do) indexes. The normalized keys are (you, are, who), (and, do, why), (you, what, say), and (you, what*, do). We mark "what" by "*" in the last key to denote that this lemma has already been taken into account by a previous key.

The Second Approach

The idea of the second approach is the following. Let query Q be a list of lemmas. In any case, we need to use the most frequently occurring lemma for a component of a three-component key. This lemma will be the first component of the first key. However, we can minimize the number of postings to read by selecting the least frequently occurring lemmas, which we can find in the query, as the other two components of the key. After we form the first key, we can apply the aforementioned logic to select the following key using the remaining lemmas of the query, etc.

When we form a key, we always need to select lemmas at different indexes in Q. For this, we will "mark" an item of the Q as "used" when we select it.

We perform the following steps in the loop.

1. If all elements of Q are "used", then we break the loop.
2. We select a lemma f with index x in the Q with the following conditions:
   a. x is not used,
   b. lemma f is the most frequently occurring lemma that satisfies the previous condition.
3. We mark x as "used".
4. We try to select a lemma s with index y in the query with the following conditions:
   a. y is not used,
   b. s is the least frequently occurring lemma that satisfies the previous condition.
5. If we cannot select a lemma in the previous step, then we select lemma s with index y in the query with the following conditions (and s is marked with * in the key):
   a. y is not equal to x.
   b. s is the least frequently occurring lemma that satisfies the previous condition.
6. We mark "y" as "used".
7. We try to select a lemma t with index z in the query with the following conditions:
   a. z is not used,
   b. t is the least frequently occurring lemma that satisfies the previous condition.
8. If we cannot select a lemma in the previous step, then we select lemma t with index z in the query with the following conditions (and t is marked with * in the key):
   a. x is not equal to z, and y is not equal to z.
   b. t is the least frequently occurring lemma that satisfies the previous condition.
9. We mark z as "used".
10. We create a three-component key (f, s, t) and include it in the list of keys.

We present two examples of the second approach.

Let us consider the subquery $SQ1$ = [who] [are] [you] [who].

With FL-numbers, we have the following: [who: 293] [are: 268] [you: 47] [who: 293].

We select "you" as the first component of the key because it is not "used" and has the most occurrence frequency in the texts, that is, the lowest FL-number of 47.

Afterwards, we select "who" as the second component of the key and "who" as the third component of the key. We have the key (you, who, who) and the normalized key (you, who, who). The indexes 0, 2 and 3 are used.

Then, we select the remaining "are" as the first component of the second key. All indexes are "used" now. Thus, we select "who" and the second "who" as the second and the third components, respectively, and we have the (are, who*, who*) key.

Let us consider the subquery $SQ2$ of another query = [who] [are] [you] [and] [why] [do] [you] [say] [what] [you] [do].

With FL-numbers we have the following: [who: 293] [are: 268] [you: 47] [and: 28] [why: 528] [do: 154] [you: 47] [say: 165] [what: 132] [you: 47] [do: 154].

We select "and: 28" as the first component of the first key and "why: 528" and "who: 293" as the second and the third, respectively. Then we select "you: 47", "are: 268", and "say: 165" for the second key. Then, we select "you: 47", "do: 154", and "do: 154" for the third key. Then we select "you: 47", "what: 132", and "why*: 528" for the last key. The normalized keys are (and, who, why), (you, say, are), (you, do, do), and (you, what, why*).

It is important to remember that we need to divide the query into parts if we have an index with a small *MaxDistance* value. Any part of the divided query must have a length that is less than or equal to the *MaxDistance*. To evaluate the subquery *SQ2* without division, we need at least a *MaxDistance* = 11.

If we consider the notes about relevance from [21], then the lengths of the parts must be less than the *MaxDistance* to some extent. For example, if the *MaxDistance* = 5, then we can limit the length of each part by the number 4, and we can consider the following division: [[who] [are] [you] [and]], [[why] [do] [you] [say]], [[what] [you] [do]]. Each of the parts must be independently evaluated, and after that, the results of these evaluations must be combined.

The Third Approach

We have an important observation regarding the second approach. When we select a high frequently occurring lemma as the first component of the key and some less frequently occurring lemma as the other component of the key, it can significantly reduce the number of postings to read. In the second approach, we select both the second and the third components of the key as the least frequently occurring lemmas. However, what if the query contains several high frequently occurring lemmas and a small number or even only two relatively low frequently occurring lemmas? In this case, it may be useful to not "spend" all of the least frequently occurring lemmas for the one key but to distribute them somehow between several keys.

In the first step, we determine the number of required keys. Let us have a subquery of length n. We need $k = n/3$ indexes, which number we need to round up. The query is a list of lemmas; thus, each item of the list has its index in the list. For any component of each key, we need to select an index of a lemma in the query. When we select an index for a specific key, we "mark" the index as "used"; thus, it cannot be used for another key.

For each key, we perform the following. We select the most frequently occurring unmarked lemma in the subquery as the first component of the key, and the least frequently occurring unmarked lemma in the subquery as the third component of the key. We perform this for all keys. Then, we need to select the second component for every key.

For each key we perform the following. If we have "unmarked" indexes, then we select the least frequently occurring unmarked lemma in the subquery as the second component of the key; otherwise, we select the least frequently occurring lemma in the subquery, whose index is not used by any component of the current key (in the latter case, the component of the key is marked with *).

Let us consider *SQ2* again. We need to define four keys.

In the first step, we define the first and the third components for each key.

We select "and: 28" as the first component of the first key and "why: 528" as the third component of the first key. We select "you: 47" and "who: 293" for the second key and "you: 47" and "are: 268" for the third key. We select "you: 47" and "say: 165" for the last key.

Then, we select "do: 154" as the second component for the first key. We select "do: 154" for the second key, "what: 132" for the third key and "why*: 528" for the last key. The normalized keys are (and, do, why), (you, do, who), (you, what, are), and (you, say, why*).

The Fourth Approach and Analysis

In the fourth approach, we consider all possible variants of key selection. In this approach, we need the ability, which we have, to estimate the count of postings for any three-component key. In this case, if we consider all possible variants of the key selection, we can select an optimal variant of the key selection (with the least number of postings to read that is required for the query evaluation).

The problem of this approach is as follows. With an increase in the length of the query, the number of variants for the key selection increases very quickly. However, this approach can be used for short queries and for analysis of optimality of other approaches.

In [15], we presented results for the first approach. In this new paper we present results for the other approaches, which are more promising.

We postpone for now the question of how to work with duplicates among the lemmas of the query.

The presented approaches can be applied not only for three-component key indexes, but also for n-component key indexes, $n > 3$, if they are needed. They can also be used in a reduced way for 2-component indexes.

3.4 Search in the Document

The algorithm of searching in the document is described in Fig. 4.

Let DID be an argument of the "Search in the document" procedure. Let us define that DID is the identifier of the current document.

The main difference between the search algorithm from [1] and the new search algorithm is described here.

For any lemma in the search query, we create an intermediate list of postings in memory. For example, let us consider the three-component index (you, are, who) and its iterator object. We create three intermediate posting lists: IL(you), IL(are), and IL(who). To fill these intermediate posting lists, we need to read postings from the (you, are, who) iterator object.

A record from the (you, are, who) iterator object has the format $(ID, P, D1, D2)$, where ID is the identifier of the document, P is the position of "you" in the document, $D1$ is the distance from "are" to "you" in the text, and $D2$ is the distance from "who" to "you" in the text.

If the lemma "are" occurs in the text after the lemma "you", then D1 > 0; otherwise, D1 < 0.

If the lemma "who" occurs in the text after the lemma "you", then D2 > 0; otherwise, D2 < 0.

We need to read from the iterator object all records with $ID = DID$.

We can produce three records from the $(ID, P, D1, D2)$ record.

We need to store the (P) record in the IL(you) intermediate posting list.

We need to store the $(P + D1)$ record in the IL(are) intermediate posting list.

We need to store the $(P + D2)$ record in the IL(who) intermediate posting list.

Let us consider the key (you, what*, do) with a lemma marked by "*". In this case, we create only two intermediate posting lists, namely, IL(you) and IL(do). The (what*) component is already taken into account by a previous key.

For each lemma of the subquery, since we have the intermediate posting list, the search is straightforward and similar to the search in the ordinary inverted file.

Additionally, an intermediate posting list is a kind of iterator object. The intermediate posting list object, IL, has the method $IL.Next$, which reads the next record from the posting list.

The intermediate posting list object, IL, has the property $IL.Value$, which contains the current record (P), where P is the position of its lemma in the document.

Let $IL.Value$ be equal to $SIZE\_MAX$ when all records are read from the IL object, where $SIZE\_MAX$ is some large number.

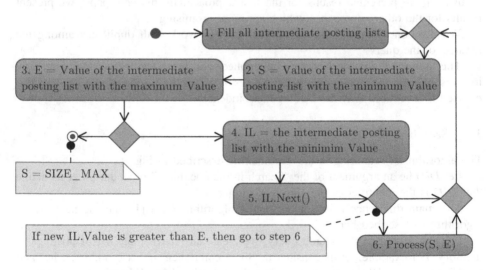

Fig. 4. UML diagram of searching in a document.

In the loop, we perform the following steps.

1. Let $MinIL$ be the intermediate posting list with a minimum value of $Value$.
2. Let $S = MinIL.Value$.
3. Let $MaxIL$ be the intermediate posting list with a maximum value of $Value$.
4. Let $E = MaxIL.Value$.
5. If there are no more records in $MinIL$, then exit from the search.

6. Execute *MinIL.Next()*.
7. If *MinIL.Value* > *E*, then execute *Process(S, E)*.
8. Go to step 1.

The *Process(S, E)* procedure adds the (*DID, S, E*) record into the result set. *S* is the position of the start of the fragment of text within the document that contains the query. *E* is the position of the end of the fragment of text within the document that contains the query.

3.5 Intermediate Posting List Data Ordering

The records (*P*) must be stored in an intermediate posting list for the given lemma in increasing order. For this requirement, the following problem arises.

Consider the text "to be or not to be or". Let the position of a word in the text be its ordinal number starting with zero. When we create the three-component key index, the following records must be stored for the key (to, be, or). The records in the format (position of "to", position of "be", position of "or") are presented below.

(to, be, or): (0, 1, 2), (0, 5, 6), (4, 1, 2), and (4, 5, 6).

From this posting list, we can create the following three intermediate posting lists.

(to): 0, 0, 4, 4; (be): 1, 5, 1, 5; and (or): 2, 6, 2, 6.

Only for the first component of the key is the intermediate posting list ordered in increasing order.

Please note that the postings in the three-component key index will actually be encoded in the (*ID, P, D1, D2*) format. For the (to, be, or) key, we will write the following posting list: (to, be, or): (*ID*, 0, 1, 2), (*ID*, 0, 5, 6), (*ID*, 4, −3, −2), and (*ID*, 4, 1, 2).

To solve the aforementioned problem, we create two binary heaps [22]. We create the first binary heap for the second component of the key. We create the second binary heap for the third component of the key.

Therefore, we will create the (be) binary heap and the (or) binary heap.

We limit the binary heap length by *MaxDistance* × 2.

When we need to read postings from the (to, be, or) posting list, we perform the following in a loop.

1. Read the next posting (*ID, P, D1, D2*) from the posting list (to, be, or).
2. Write (*P*) into the (to) intermediate posting list.
3. Write (*P* + *D1*) into the (be) binary heap.
4. Let *M* be the first (the minimum element) of the (be) binary heap. If the length of the (be) binary heap is greater than *MaxDistance* × 2 or if the distance between *M* and the new element (*P* + *D1*) is greater than *MaxDistance* × 2, then remove the first element from this binary heap, and write it into the (be) intermediate posting list.
5. Write (*P* + *D2*) into the (or) binary heap.
6. Let *M* be the first (the minimum element) of the (or) binary heap. If the length of the (or) binary heap is greater than *MaxDistance* × 2 or if the distance between *M* and the new element (*P* + *D2*) is greater than *MaxDistance* × 2, then remove the first element from this binary heap, and write it into the (or) intermediate posting list.
7. Go to step 1.

Let us consider a key (f, s, t) and its posting list, L. We create three intermediate posting lists and two binary heaps to proceed as follows.

1. Intermediate posting list F for f.
2. Intermediate posting list S and binary heap SH for s.
3. Intermediate posting list T and binary heap TH for t.

Let us introduce the methods *PopMin, Min* and *Length* of a binary heap object. The *PopMin* method returns the minimum element from the binary heap and removes this element from the binary heap. The *Length* method returns the length of the binary heap. The *Min* method returns the minimum element from the binary heap but does not change the binary heap.

In Fig. 5, we present the UML diagram of the posting list L reading process.

After all postings from L are read, we need to write all elements from the binary heaps to their intermediate posting lists.

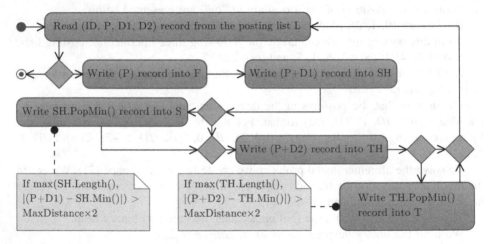

Fig. 5. UML diagram of the posting list reading process.

3.6   Advantages of the New Algorithm

The new algorithm may require a smaller amount of posting lists to evaluate a search query in comparison with the algorithm from [1] and, therefore, provides faster searches. It also allows for a more flexible key selection.

3.7   Computational Complexity

Let Q be a subquery of m lemmas. Let n be the total number of postings to read for the query evaluation.

For each posting, we need to use it in the *Equalize* procedure. In [1], the author states that the cost of such usage is $O(\log(m))$.

For each posting, we need to add it to the three intermediate posting lists. The cost of this process is $O(\log(MaxDistance))$ (see Sect. 3.5).

For each posting, we need to use it when searching in a document. The cost of this process is $O(\log(m))$ (see Sect. 3.4).

The final cost of the subquery evaluation is

$$O(n \cdot (\log(m) + \log(MaxDistance)) = O(n \cdot \log(\max(m, MaxDistance))).$$

4 Search Experiments

4.1 Search Experiment Environment

All search experiments were conducted using a collection of texts from [1]. The total size of the text collection was 71.5 GB. The text collection consisted of 195 000 documents of plain text, fiction and magazine articles. We used $MaxDistance = 5$, $SWCount = 700$, and $FUCount = 2100$. The search experiments were conducted using the experimental methodology from [1].

We assume that in typical texts, words are distributed similarly, in accordance with Zipf's law [6]. Therefore, the results obtained with our text collection will be relevant to other collections.

We used the following computational resources:

CPU: Intel(R) Core(TM) i7 CPU 920 @ 2.67 GHz.
HDD: 7200 RPM. RAM: 24 GB.
OS: Microsoft Windows 2008 R2 Enterprise.

We created the following indexes.

Idx1: the ordinary inverted index without any improvements, such as NSW records [1, 19]. The total size was 95 GB.
Idx2: our indexes, including the ordinary inverted index with the NSW records and the (w, v) and (f, s, t) indexes, where $MaxDistance = 5$. The total size was 746 GB.

Please note that the total size of each type of index includes the size of the repository (indexed texts in compressed form), which was 47.2 GB.

4.2 Search Results

There are 975 queries, and all queries consisted only of stop lemmas. The query set was selected as in [1]. All searches were performed in a single program thread. We searched all queries from the query set with different types of indexes to estimate the performance gains of our indexes. The query length was from 3 to 5 words.

Studies by Jansen et al. [23] have shown that queries with lengths greater than 5 are very rare. In [23], query logs of a search system were analyzed, and it was established that queries with a length of 6 represent approximately 1% of all queries and that fewer than 4% of all queries had more than 6 terms.

We performed the following experiments.

SE1: all queries are evaluated using the standard inverted index Idx1.

SE2.1: all queries are evaluated using Idx2 and the algorithm from [1].

SE2.2: all queries are evaluated using Idx2, the novel algorithm presented in this paper and that in [15] with the key selection based on the first approach.

SE2.3: all queries are evaluated using Idx2 and the novel algorithm presented in this paper with the key selection based on the second approach.

SE2.4: all queries are evaluated using Idx2 and the novel algorithm presented in this paper with the key selection based on the third approach.

SE2.5: all queries are evaluated using Idx2 and the novel algorithm presented in this paper with the key selection based on the fourth approach.

Average query times:

SE1: 31.27 s, SE2.1: 0.33 s, SE2.2: 0.29 s, SE2.3: 0.24 s, SE2.4: 0.24 s, and SE2.5: 0.27 s.

Average data read sizes per query:

SE1: 745 MB, SE2.1: 8.45 MB, SE2.2: 6.82 MB, SE2.3: 6.2 MB, SE2.4: 6.16 MB, and SE2.5: 5.79 MB.

Average numbers of postings per query:

SE1: 193 million, SE2.1: 765 thousand, SE2.2: 559 thousand, SE2.3: 423 thousand, SE2.4: 419 thousand, and SE2.5: 411 thousand.

We improved the query processing time by a factor of 94.7 with the SE2.1 algorithm, by a factor of 107.8 with the SE2.2 algorithm, and by a factor of 130 with the SE2.3 and SE2.4 algorithms (see Fig. 6).

In SE2.5 we observed a slight increase in the average query execution time, because this time includes checking all the possible combinations of the three-component key selection. We used the SE2.5 results only for analysis of the effectiveness of SE2.3, and SE2.4 in relation to SE2.2 (see later); therefore, SE2.5 is excluded from Fig. 6.

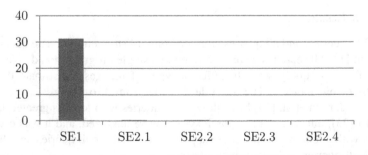

Fig. 6. Average query execution times for SE1, SE2.1, SE2.2, SE2.3, and SE2.4 (in seconds).

The left-hand bar shows the average query execution time with the standard inverted indexes. The subsequent bars show the average query execution times with our indexes using the SE2.1, SE2.2, SE2.3 and SE2.4 algorithms. Our bars are much smaller than the left-hand bar because our searches are very quick.

We improved the data read size per query by a factor of 88 with SE2.1, by a factor of 109.2 with SE2.2 and by a factor of 120 with SE2.3 and SE2.4 (see Fig. 7).

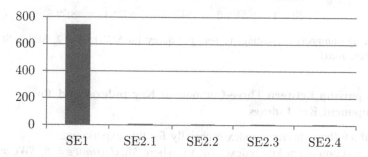

Fig. 7. Average data read sizes per query for SE1, SE2.1, SE2.2, SE2.3 and SE2.4 (MB).

The left-hand bar shows the average data read size per query for SE1. The subsequent bars show the average data read size per query for SE2.1, SE2.2, SE2.3 and SE2.4.

We show how SE2.3 and SE2.4 outperform SE2.2 in Fig. 8.

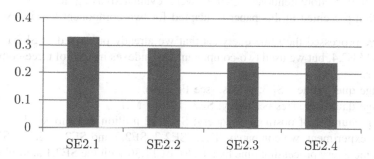

Fig. 8. Average query execution times for SE2.1, SE2.2, SE2.3 and SE2.4 (in seconds).

We show the average number of postings to read per query for SE2.1, SE2.2, SE2.3, SE2.4 and SE2.5 in Fig. 9. We observe that SE2.3 and SE2.4 have similar effectiveness in comparison with SE2.5, which is the optimal key selection. We also observe how SE2.3 and SE2.4 outperform the original SE2.2 method.

SE2.3 and SE.2.4 have equal performance on average; however, we have examples of queries that have significantly different execution times for the SE2.3 and SE2.4 approaches. If we have information about the posting list length for every key, then it will be good to quickly check both SE2.3 and SE2.4 strategies before evaluating a specific query.

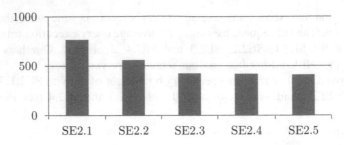

Fig. 9. Average numbers of postings to read per query for SE2.1, SE2.2, SE2.3, SE2.4 and SE2.5 (in thousands).

4.3 Comparison Between Three-Component Key Indexes and Two-Component Key Indexes

We created another additional index especially for this experiment.

Idx3: two-component key indexes (w, v), where *MaxDistance = 5, SWCount = 0,* and *FUCount = 700.* The total index size is 275 GB.

In this case, for any two lemmas, w and v, where $w \leq v$, $FL(w) < 700$, and $FL(v) < 700$, we have a two-component key index (w, v).

Each posting in this index includes the distance between w and v in the text.

Such w and v lemmas are stop lemmas for Idx2.

We performed the following experiment:

SE3: all 975 aforementioned queries were evaluated using Idx3, and the new algorithm presented in this paper is adapted for two-component key indexes.

In SE3, we processed the same query set that we already processed in SE2.1, SE2.2, SE2.3, and SE2.4, but we used two-component key indexes instead of three-component key indexes.

Average query times: SE3: 3.75 s. (see Fig. 10).

Average data read sizes per query: SE3: 105.17 MB.

Average number of postings per query: SE3: 12 million 761 thousand.

In this experiment, we compared SE2.1, SE2.2, SE2.3 and SE2.4 against SE3. We improved the query processing time by a factor of 11.36 with the SE2.1 algorithm, by a factor of 12.93 with the SE2.2 algorithm, and by a factor of 15.6 with SE2.3 and SE2.4 in comparison with the two-component key index (SE3) case (see Fig. 10).

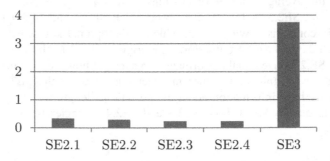

Fig. 10. Average query execution times for SE2.1, SE2.2, SE2.3, SE2.4 and SE3 (in seconds).

The left-hand bar shows the average query execution time with the three-component key indexes using the algorithm from [1]. The three center bars show the average query execution time with the three-component key indexes using the new algorithm described in this paper. The right-hand bar shows the average query execution time with the two-component key indexes.

The bars that related to the three-component key indexes are much smaller than the right-hand bar because the three-component key indexes enable much quicker searches than the two-component key indexes.

This experiment shows that three-component key indexes **by an order of magnitude are more effective** than the two-component indexes when the queries that consist of stop lemmas are evaluated.

We improved the data read size per query by a factor of 12.44 with SE2.1, by a factor of 15.42 with SE2.2 and by a factor of 16.96 with SE2.3 and SE2.4 in comparison with the two-component key index (SE3) case (see Fig. 11).

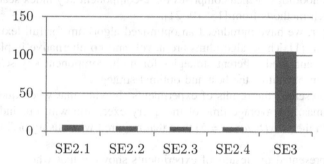

Fig. 11. Average data read sizes per query for SE2.1, SE2.2, SE2.3, SE2.4 and SE3 (MB).

The left-hand bar shows the average data read size per query with SE2.1. The subsequent bars show the average data read size per query with SE2.2, SE2.3, SE2.4 and SE3.

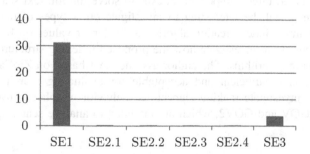

Fig. 12. Average query execution times for SE1, SE2.1, SE2.2, SE2.3, SE2.4 and SE3 (in seconds).

We show the average query execution time for all experiments in Fig. 12.

The left-hand bar shows the average query execution time with the standard inverted indexes. The four subsequent bars show the average query execution times with the three-component key indexes for the SE2.1, SE2.2, SE2.3 and SE2.4 algorithms. The right-hand bar shows the average query execution time with the two-component key indexes in the SE3 experiment.

5   Conclusion and Future Work

A query that contains high-frequency occurring words induces performance problems. To solve these performance problems and to satisfy the fastidious demands of the users, we developed and elaborated three-component key indexes.

In this paper, we investigated searches with queries that contain only stop lemmas. Other query types were studied in [18, 19, 21]. As we discussed in [1], three-component key indexes are an important and integral part of our comprehensive full-text search methodology, which comprises three-component key index search methods and other search methods from [18, 19, 21].

In this paper, we have introduced an optimized algorithm for full-text searches in comparison with [1]. These algorithms are novel, and no alternative implementations exist. We have analyzed different strategies for multi-component key selection for a specific query in pursuit of the best and optimal strategy.

We have presented the results of experiments showing that when queries contain only stop lemmas, the average time of the query execution with our indexes is 130 times less (with the *MaxDistance* = 5) than that required when using ordinary inverted indexes.

We have presented the results of experiments showing that when queries contain only stop lemmas, the average time of the query execution with our indexes is 15.6 times less (with the *MaxDistance* = 5) than that required when using two-component key indexes.

Using the last experiment, we diligently prove that three-component indexes are stupendous and cannot be replaced by two-component key indexes. This is the reason why we implemented three-component indexes to solve the full-text search task.

In the future, it will be interesting to investigate other types of queries in more detail and to optimize index creation algorithms for larger values of *MaxDistance*. It will also be important to investigate how the proposed indexing structure can be used by modern ranking algorithms. The author assumes that based on Zipf's law [6], our test text collection is sufficient and acceptable for evaluating search performance. Nevertheless, to investigate ranking algorithms' behavior we plan to use collections, such as TREC GOV and GOV2, which are intended to analyze search quality.

References

1. Veretennikov, A.B.: Proximity full-text search with response time guarantee by means of three component keys. Bull. South Ural State Univ. Ser: Comput. Math. Softw. Eng. **7**(1), 60–77 (2018). https://doi.org/10.14529/cmse180105. (in Russian)
2. Buttcher, S., Clarke, C., Lushman, B.: Term proximity scoring for ad-hoc retrieval on very large text collections. In: SIGIR 2006, pp. 621–622 (2006). https://doi.org/10.1145/1148170.1148285
3. Rasolofo, Y., Savoy, J.: Term proximity scoring for keyword-based retrieval systems. In: European Conference on Information Retrieval (ECIR) 2003: Advances in Information Retrieval, pp. 207–218 (2003). https://doi.org/10.1007/3-540-36618-0_15
4. Schenkel, R., Broschart, A., Hwang, S., Theobald, M., Weikum, G.: Efficient text proximity search. In: Ziviani, N., Baeza-Yates, R. (eds.) SPIRE 2007. LNCS, vol. 4726, pp. 287–299. Springer, Heidelberg (2007). https://doi.org/10.1007/978-3-540-75530-2_26
5. Yan, H., Shi, S., Zhang, F., Suel, T., Wen, J.-R.: Efficient term proximity search with term-pair indexes. In: CIKM 2010 Proceedings of the 19th ACM International Conference on Information and Knowledge Management, Toronto, ON, Canada, 26–30 October 2010, pp. 1229–1238 (2010). https://doi.org/10.1145/1871437.1871593
6. Zipf, G.: Relative frequency as a determinant of phonetic change. Harv. Stud. Class. Philol. **40**, 1–95 (1929). https://doi.org/10.2307/408772
7. Luk, R.W.P.: Scalable, statistical storage allocation for extensible inverted file construction. J. Syst. Softw. **84**(7), 1082–1088 (2011). https://doi.org/10.1016/j.jss.2011.01.049
8. Tomasic, A., Garcia-Molina, H., Shoens, K.: Incremental updates of inverted lists for text document retrieval. In: SIGMOD 1994 Proceedings of the 1994 ACM SIGMOD International Conference on Management of Data, Minneapolis, Minnesota, 24–27 May 1994, pp. 289–300 (1994). https://doi.org/10.1145/191839.191896
9. Zobel, J., Moffat, A.: Inverted files for text search engines. ACM Comput. Surv. **38**(2), Article no. 6 (2006). https://doi.org/10.1145/1132956.1132959
10. Miller, R.B.: Response time in man-computer conversational transactions. In: Proceedings: AFIPS Fall Joint Computer Conference. San Francisco, California, 09–11 December 1968, vol. 33, pp. 267–277 (1968). https://doi.org/10.1145/1476589.1476628
11. Anh, V.N., de Kretser, O., Moffat, A.: Vector-space ranking with effective early termination. In: SIGIR 2001 Proceedings of the 24th Annual International ACM SIGIR Conference on Research and Development in Information Retrieval, New Orleans, Louisiana, USA, 9–12 September 2001, pp. 35–42 (2001). https://doi.org/10.1145/383952.383957
12. Garcia, S., Williams, H.E., Cannane, A.: Access-ordered indexes. In: ACSC 2004 Proceedings of the 27th Australasian Conference on Computer Science, Dunedin, New Zealand, 18–22 January 2004, pp. 7–14 (2004)
13. Bahle, D., Williams, H.E., Zobel, J.: Efficient phrase querying with an auxiliary index. In: SIGIR 2002 Proceedings of the 25th Annual International ACM SIGIR Conference on Research and Development in Information Retrieval, Tampere, Finland, 11–15 August 2002, pp. 215–221 (2002). https://doi.org/10.1145/564376.564415
14. Williams, H.E., Zobel, J., Bahle, D.: Fast phrase querying with combined indexes. ACM Trans. Inf. Syst. (TOIS) **22**(4), 573–594 (2004). https://doi.org/10.1145/1028099.1028102
15. Veretennikov, A.B.: Proximity full-text search with a response time guarantee by means of additional indexes with multi-component keys. In: Selected Papers of the XX International Conference on Data Analytics and Management in Data Intensive Domains (DAMDID/RCDL 2018), Moscow, Russia, 9–12 October 2018, pp. 123–130 (2018). http://ceur-ws.org/Vol-2277

16. Veretennikov, A.B.: O poiske fraz i naborov slov v polnotekstovom indekse (About phrases search in full-text index). Control Syst. Inf. Technol. **48**(2.1), 125–130 (2012). (in Russian)
17. Veretennikov, A.B.: Effektivnyi polnotekstovyi poisk s uchetom blizosti slov pri pomoshchi trekhkomponentnykh klyuchei (Efficient full-text proximity search by means of three component keys). Control Syst. Inf. Technol. **69**(3), 25–32 (2017). (in Russian)
18. Veretennikov, A.B.: Ispol'zovanie dopolnitel'nykh indeksov dlya bolee bystrogo polnotekstovogo poiska fraz, vklyuchayushchikh chasto vstrechayushchiesya slova (Using additional indexes for fast full-text searching phrases that contains frequently used words). Control Syst. Inf. Technol. **52**(2), 61–66 (2013). (in Russian)
19. Veretennikov, A.B.: Effektivnyi polnotekstovyi poisk s ispol'zovaniem dopolnitel'nykh indeksov chasto vstrechayushchikhsya slov (Efficient full-text search by means of additional indexes of frequently used words). Control Syst. Inf. Technol. **66**(4), 52–60 (2016). (in Russian)
20. Veretennikov, A.B.: Sozdanie dopolnitel'nykh indeksov dlya bolee bystrogo polnotekstovogo poiska fraz, vklyuchayushchikh chasto vstrechayushchiesya slova (Creating additional indexes for fast full-text searching phrases that contains frequently used words). Control Syst. Inf. Technol. **63**(1), 27–33 (2016). (in Russian)
21. Veretennikov, A.B.: Proximity full-text search with a response time guarantee by means of additional indexes. In: Arai, K., Kapoor, S., Bhatia, R. (eds.) IntelliSys 2018. AISC, vol. 868, pp. 936–954. Springer, Cham (2019). https://doi.org/10.1007/978-3-030-01054-6_66
22. Williams, J.W.J.: Algorithm 232 – Heapsort. Commun. ACM **7**(6), 347–348 (1964)
23. Jansen, B.J., Spink, A., Saracevic, T.: Real life, real users and real needs: a study and analysis of user queries on the Web. Inf. Process. Manag. **36**(2), 207–227 (2000). https://doi.org/10.1016/S0306-4573(99)00056-4

Scope and Challenges of Language Modelling - An Interrogative Survey on Context and Embeddings

Matthias Nitsche and Marina Tropmann-Frick[✉]

Department of Computer Science, Hamburg University of Applied Sciences,
Hamburg, Germany
{matthias.nitsche,marina.tropmann-frick}@haw-hamburg.de

Abstract. In this work we explore the domain of Language Modelling. We focus here on different context selection strategies, data augmentation techniques, and word embedding models. Many of the existing approaches are difficult to understand without specific expertise in this domain. Therefore, we concentrate on appropriate explanations and representations that enable us to compare several approaches.

Keywords: Natural language processing · Neural language model · Embeddings · Context selection · Data augmentation

1 Introduction

Language is a high dimensional and multi-sense problem domain dealing with polysemy, synonymy, antonymy, hyponymy etc. The language processing pipeline often starts with morphological treatment of text, e.g. stemming, stopword removal, special character extraction and tokenization. The next step is usually a projection of words for text representation. Classical models project words using WordNet mapping each word to a relation, employ methods from linear algebra like Singular Value Decomposition (SVD) and most famously Latent Semantic Indexing (LSI) [8]. More complicated statistical models involve expectation maximization procedures for which Latent Dirichlet Allocation (LDA) [4] is the standard.

Word and subword-level embeddings try to overcome some of the limitations of the former methods using neural networks posing language models as an optimization problem. Word2Vec by [22] was the first successful model that superseded the quality of preceding methods. Embeddings map words, sentences, characters or part of words to a non-linear latent space in $\mathbb{R}^l$ where l stands for the amount of dimensions the embedding has. Due to corporations like Google and Facebook that push forward research in the area of machine learning and deep learning, the transfer of embedding models have become easier. Projects like *fastText, spaCy, Starspace, GloVe* and *Word2Vec Googles News embeddings* offer pre-trained language models on vast amounts of data.

© Springer Nature Switzerland AG 2019
Y. Manolopoulos and S. Stupnikov (Eds.): DAMDID/RCDL 2018, CCIS 1003, pp. 131–145, 2019.
https://doi.org/10.1007/978-3-030-23584-0_8

Most real world language datasets are considerably smaller than those of such corporation as Facebook or Google and thus suffer greatly with out-of-vocabulary (OOV) words. Further neural networks need vast amounts of data to overcome problems with overfitting. Thus generalization of language models needs attention and techniques like domain adaptation or transfer learning try to overcome this gap.

This work represents an extension of our previous work in [27]. Several modifications and additions are done to different parts of the previous paper. Furthermore, the last section of this work "New Directions" presents new state-of-art approaches in the literature. These rely heavily on a combination of two ideas - attention-based modelling [35] and unsupervised pre-training [31]. Due to significant improvements in computational efficiency and model performance, these approaches will dominate the future research in the area of Natural Language Processing.

2   Stochastic Language Modelling

The domain of language modelling experienced large changes since the starting period of natural language processing in the 1960s [7]. Early symbolic approaches made an attempt to capture the meaning of text using rules written by humans developing rule-based systems. Such systems were limited to particular domains they were designed for [38] and unable to deal with unseen or unexpected inputs.

Over the last 20 years - since the late 1990s [17,20,25], a statistical approach of natural language processing has become commonplace and only slowly gives way to 'new' deep neural network-based approaches in recent years.

The main task of a stochastic language model is to provide estimates of the probability of a word sequence. This probability is the result of joint computation of conditional probabilities of words in a sentence or sequence generally using the principle of maximum likelihood estimation. This problem is usually reduced to learning the conditional distribution of the next word given a fixed number of preceding words. This task can be accomplished successfully with n-gram models. The starting point here were unigram models, following by the bigram-, trigram-, etc. models. Because language has rather long-distance dependencies, the n-gram models are the most efficient here. For learning tasks a distinction can be made between discriminative models that model the conditional probability directly from raw data and generative models that learns the joint probability distribution.

The area of statistical language modelling remains still an interesting topic of research [24] although machine learning methods based on neural networks have become the main tools in natural language processing. Also extensions of the discussed approaches can be found e.g. in [37], where the n-grams are presented in form of *charagrams* - an approach to learn character-level compositions. We describe this approach more detailed in the following sections.

3   Character-Level Embeddings

Character-level embeddings deal with words by slicing them into smaller proportions. This is advantageous due to the fact that single words and their corresponding vectors only match by symbolic comparison. Thus there are advantages of representing words as vectors of sub-level symbolic representations, that first largely occurred in neural machine translations. The representations range from character CNNs/LSTMs [16] to character n-grams [5,37].

Character-level embedding models typically build on pre-trained word embeddings. Additionally character based representations of words are itself vectors for each character of a word or vector representation of the n-grams of a word. [15] explore different architectures for language modelling and compared three different models with differing inputs to language models.

[16] presents a model with a character-level convolutional neural network (CNN) with a highway network over characters. Characters are used as an input to a single layer CNN with max-pooling, using a highway network, introduced in [33], similarly to a RNN with a carry mechanism, before applying a LSTM with a softmax for the most likely next word representation. Most interesting in this work is the application of the CNN with the highway network.

The vocabulary C over characters and d as usual the embedding size, we deal with $\mathbb{R}^{d \times |C|}$ matrix character embeddings. A word $k \in V$ is decomposed as a sequence of characters $[c_1, \ldots, c_l]$, where $l = |k|$, the matrix representation then is $C^k \in \mathbb{R}^{d \times l}$. The columns are character vectors, the rows character dimensions d. The character-level CNN maximizes the following cost function

$$f^k[i] = \tanh(\langle C^k[*, i : i + w - 1], \mathrm{H}\rangle + b) \tag{1}$$
$$y^k = \max_i(f^k[i]) \tag{2}$$

where C^k is a filter of width w creating a feature map f^k, indexed by $i \ldots i+w-1$ columns over the filters of C^k. $\langle \ldots \rangle$ is the inner product. The convolution or kernel can be seen as a generator for character n-grams. This is then fed to y^k which takes the maximum of the feature map, e.g., applies a max pooling transformation. After this y^k is used as input to a highway network, which is essentially a RNN/LSTM network with different gating mechanisms.

$$z = t \odot g(W_H \cdot y + b_H) + (1 - t) \odot y \qquad Memory\,transformation \tag{3}$$
$$t = \sigma(W_T \cdot y + b_T) \qquad Transform\,gate \tag{4}$$

The transform gate t maps the input into a different latent space, $(1-t)$ is the carry gate, deciding what information will carry on over time. $g(W_H y + b_H$ is a typical affine transformation with a non-linearity applied. $\odot$ is the entry-wise product or Hadamard Product. Stacking several layers of highway networks allow to carry parts of the input to the output, while combining them in a recurrent fashion. At last the output z is fit into an LSTM with a softmax to obtain distributions over the next word. [16] manages to reduce parameter size by 60% while achieving state of the art language modelling results. Furthermore they find that their models learn semantic and orthographic relations from characters arguing if word-level embeddings seem even necessary. They also successfully deal with Out-of-vocabulary words (OOV) assigning intrinsically chosen words like *looooook* to the correct word *look*, that word-level models failed to learn.

3.1 Character n-grams

While character-level models work on par with word-level models, recent works focused on character n-grams. *Charagram* by [37] is an approach to learn character-level compositions, not the statistics of single characters as we described above. Given a textual word or sentence e.g. a sequence of characters x

$$x = \langle x_1, x_2, \ldots, x_m \rangle, \qquad Character\text{-}level\ textual\ sequence \qquad (5)$$

$$x_j^i = \langle x_i, x_{i+1}, \ldots, x_j \rangle \qquad Sub\text{-}sequence\ of\ characters\ from\ i\ to\ j \qquad (6)$$

Charagram produces a character n-gram count vector, where each character n-gram has its own vector $W^{x_j^i}$, if the n-gram $x_j^i \in V$ is part of all n-grams of the model. f is the indicator function, if $x_j^i \in V$ then 1 else 0.

$$g_{\text{char}}(x) = h(b + \sum_{i=1}^{m+1} \sum_{j=1+i-k}^{i} f(x_j^i \in V) W^{x_j^i}) \qquad (7)$$

h here is a single non-linearity applied over the sum of all n-gram character vectors of x, where k is the maximum length of any character n-gram in the model. V can be initialized by different choices as a model parameter.

It was also found that the models could be trained on far fewer examples while still being comparable to state-of-the-art models. OOV (out-of-vocabulary) words are handled naturally, because *Charagram* represents words as the sum of characters that even unseen words can be trained and successfully embedded. [37] have shown that character n-grams can be used to beat state of the art models trained on words. [5] proposed an architecture using the Word2Vec skip-gram objective on a bag of words of character n-grams. The authors describe the skip-gram with negative sampling introduced by [22] and exchange the respective scoring function. Word2Vec takes two vectors u_w and v_w element in $\mathbb{R}^d$, where d is the dimensionality and u_{w_t} is the target word vector with the corresponding context vectors v_{w_c}

$$s(w_t, w_c) = u_{w_t}^T \cdot v_{w_c} \qquad Word\text{-}level\ objective \qquad (8)$$

We would like to represent a word as a character representation through n-grams, e.g., where $= \langle wh, whe, her, ere, re \rangle$. The above Word2Vec objective can be rewritten to represent each word as a bag of character n-grams vector representation

$$s(w, c) = \sum_{g \in \mathbb{G}_w} z_g^T v_c \qquad Character\text{-}level\ objective \qquad (9)$$

where z_g is a vector representation of a single n-gram, from a global set $\mathbb{G}$ with all character n-grams. We are interested in the Word2Vec objective where each word is now a sum of these character n-gram representations $\mathbb{G}_w \subset 1, \ldots, \mathbb{G}$. [5] successfully improve on the analogy task over previous models and deal with OOV words even where the morphemes do not match up. The size of n-grams matter and they suggest above $n > 2$ or $n \geq 5$ for languages like German with many noun compounds.

4   Word Embeddings

At first we will briefly review word-level embeddings. Corpora typically consist of words that are part of sentences in documents. Before embeddings can be trained each sentence is tokenized and morphologically altered with stemming or lemmatization.

4.1   Bag of Words

Many classical language models represent a text as a bag of words model, where words are represented as a co-occurrence feature matrix. Each entry corresponds to the number of occurrences of some word from the given vocabulary in the text. This word frequency can be represented weighted with term frequency-inverse document frequency (tf-idf), which additionally reflects to importance of a term is in a corpus, our given collection of texts. The bag-of-words ignores grammar and word order.

4.2   Out-of-Vocabulary Words

Out-of-vocabulary words (OOV) is a problem in two circumstances. The first is that the amount of OOV words is large and second - the dataset is small and deals with niche words where every word constitutes heavily. Words that do not match any given word vector are mapped to the *UNK* token. There are several strategies on dealing with OOV words ranging from using the context words around OOV words [13], using pre-trained language models to assign their vector to OOV words [9] or retrain character-level language models on pre-trained models [30]. [13] suggests a few tricks to improve on Word2Vec with their proposed model Nonce2Vec. They use pre-trained word embeddings from Word2Vec and treat OOV words as the sum of their context words. They show that this is applicable on smaller datasets as well.

[9] found it effective to use vectors of pre-trained language models where a word was OOV in their domain. Using the pre-trained vector of a different domain helped them in improving the initialization of their OOV words in comparison to assign a global *UNK* token to their data points. They improved models on reading comprehension considerably especially with OOV words. [30] have shown that generating OOV word embeddings by training a character-level model on a pre-trained dataset. The goal is to re-create the vectors by leveraging

character information. With a character-level vector word representation OOV words can be handled based on the sum of character vectors. They have found that this is much better in cases where the dataset is small and pre-trained embeddings are available.

4.3   Word2Vec

[22] improved on several aspects of Bengio's model by using the skip-gram window function (an alternative would be CBOW) and a tractable approximation of the softmax called negative sampling/hierarchical softmax. Word2Vec has become the de facto standard in a lot of language downstream tasks. Google shipped pre-trained Word2Vec skip-gram models on Google News articles for everybody to use. The corpus is large (up to a billion words) and the dimensions of the latent space is large $d = 300$. The training would take weeks up to months on a just a few state-of-the-art GPUs, saving each researcher the time to train them themselves. We will see a great influx of pre-trained language models in the future because OOV words are a real issue and generalization on small sparse domains is highly problematic. While most of the premises of pre-trained models are great, they also introduce biases. [6] have shown that this particular dataset employs gender biases. Figure 1 shows the skip-gram objective on the left side and CBOW on the right. *Skip-gram* predicts the context of a center word w_i over a window c such that $w_{i-c}, \dots, w_i, \dots w_{i+c}$ is satisfied.

$$\frac{1}{T} \sum_{t=1}^{T} \sum_{-c \le j \le c, j \ne 0} \log p(w_{t+j}|w_t) \tag{10}$$

CBOW does the opposite, given a word context $w_{i-c}, \dots, w_i, \dots w_{i+c}$ predict the center word w_i that is most likely. Negative sampling speeds up the

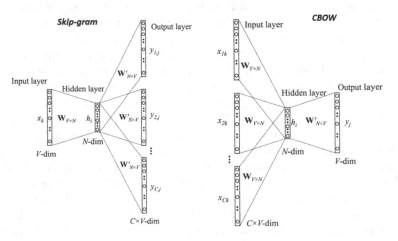

Fig. 1. The Skip-gram Model and Continuous bag-of-word model as shown in [32]

performance by using the positive samples of the context words $2 * c$ and uses only a few negative samples that are not in its context. The respective objective cost function is

$$\log\sigma(\mathrm{v}'^{T}_{w_O} v_{w_I}) + \sum_{t=1}^{k} \mathbb{E}_{w_i} \sim P_n(w)[\log\sigma(-\mathrm{v}'^{T}_{w_O} v_{w_I})] \tag{11}$$

where σ is the sigmoid function, a binary function, drawing k samples from the negative or noise distribution $P_n(w)$, to distinguish the negative draws $\mathbb{E}_{w_i}$ from the target word w_O drawn from the context of w_I. The objective of negative sampling is to learn high quality word embeddings by comparing noise (out of context) to words from the context. For an in-depth review of negative sampling and noise contrastive estimation (NCE) see [10].

Another language model building upon Word2Vec is Global Vectors for Word Representation (*GloVe*) by [28], which is trained on an aggregated global word co-occurrence matrix from a corpus. The difference to Word2Vec is that the global statistics are taken into account contrary to Word2Vec, that works on local context windows alone. GloVe typically performs better than Word2Vec skip-gram, especially when the vocabulary is large. *GloVe* is also available on different corpora such as Twitter, Common Crawl or Wikipedia. For practitioners: it is suggested to use *GloVe* wherever applicable or usable.

4.4 Bag of Tricks - FastText

Another interesting and popular word embedding model is *fastText* by [14]. It bases on a similar idea as Word2Vec. Instead of negative sampling - using the hierarchical softmax, and instead of words - using n-gram features. N-grams build on bag of words, commonly known as a co-occurrence matrix $D \times V$ where documents D are rows and the whole vocabulary V the features assuming i.i.d word order. Given a sequence of words $[w_1, \ldots, w_k]$ n-grams take slices of n e.g. $[[w_1, \ldots, w_n]_1, \ldots, [w_{i+1}, \ldots, w_{n+1}]_k]$. *fastText* comes in two flavours: character-level and word-level n-grams. We will review the character-level n-grams later.

$$-\frac{1}{N} \sum_{n=1}^{N} y_n \log(f(BAx_n)) \tag{12}$$

Formula 12 is the corresponding cost function, where f is the hierarchical softmax function, x_n is a document with bag of n-gram feature vectors, A and B are weight matrices and y_n the label given a classification task. The label y in this case is the word. The unsupervised learning task is the hierarchical softmax with $CBOW$ denoted as f and has the following form:

$$P(y = j|C) = \frac{exp(\beta_j^T \cdot C)}{\sum_{t=1}^{|V|} exp(\beta_t^T \cdot C)}. \tag{13}$$

As can be seen instead of finding the surrounding context of a word w we try to find the most probable word given the context C. What is novel about this approach is using n-gram features instead of windows speeding up the training, while still matching state-of-the-art results. *fastText* training time on a sentiment analysis task was 10 s compared to the shortest running model of 2–3 h up to several days. As we will see later, this model can be largely improved with character n-grams proposed in [5] and [37].

4.5 CoVe

So far we have investigated shallow neural networks with single layers and therefore only one non-linearity. [21] have found that training an attentional sequence-to-sequence model normally used for neural machine translations helps at enriching word vectors not just on the word-level hierarchy. By training a two-layer, bidirectional long short-term memory [12], on a source language (English) to a target (German) they achieve state-of-the-art performance. All sequences of words w^x are pre-initialized with $GloVe(w^x)$ where words become sequences of vectors.

$$w^x = [w_1^x, \dots, w_n^x] \tag{14}$$

$$GloVe(w^x) = [GloVe(w_1^x), \dots, GloVe(w_n^x)] \tag{15}$$

$$w^z = [w_1^z, \dots, w_n^z] \qquad Randomly\,initialized \tag{16}$$

where w^x is a sentence in the source language and w^z of the target language maximizing the likelihood of an encoder MT-LSTM h, a decoder LSTM h_t^{dec}.

$$h = MT\text{-}LSTM(GloVe(w^x)) \qquad\qquad Encoder \tag{17}$$

$$h_t^{dec} = LSTM([z_{t-1}; \widetilde{h}_{t-1}], h_{t-1}^{dec}) \qquad\qquad Decoder \tag{18}$$

$$\alpha_t = softmax(H(W_1 h_t^{dec} + b_1)) \qquad\qquad Attention \tag{19}$$

$$\widetilde{h}_t = [\tanh(W_2 H^T \alpha_t + b_2); h_t^{dec}] \qquad Concatenated\,attentional\,sum \tag{20}$$

$$y = softmax(W_{out}\widetilde{h}_t + b_{out}) \qquad Output\,word\,distribution \tag{21}$$

The softmax attention α_t over the decoder h_t^{dec} represents the relevance of each step from the encoder h. $\widetilde{h}_t$ then is a hidden state where the softmax and h_t^{dec} are concatenated, possibly to attend to the relevant parts while not forgetting what was learned during the decoding. Intuitively we are training a machine translation model where the only interesting part are the learned context vectors for sequences of the *MT-LSTM*.

It was shown that the model performs better by concatenating GloVe and CoVe into one single vector. The idea behind this is that we can transfer the higher level features learned in sequence-to-sequence tasks to standard downstream tasks like classification. By first using GloVe on the word-level and then the MT-LSTM we are creating layers of abstractions. Essentially this is a first step towards transfer learning, which is standard practice in computer vision tasks with pre-trained CNNs. The top achiever is a model called Char + CoVe-L with a large *CoVe* model concatenated with a *n-gram* character features model introduced by [14] in fastText.

4.6 Dict2Vec

Word2Vec, GloVe and fastText create strong baseline models for word embeddings. Newer trends also incorporate additional information from external data sources, augmenting word vectors. [34] improve on the Word2Vec model by [22] using dictionaries. The key concept presented in [34] is that each word can be weakly and strongly linked to each other given the definition.

For instance the *Guitar* and *Violin* share the words *stringed musical instrument*, that should strongly tie them together. In the definition of the *Violin* there is no *plucking* or *strumming* and thus is considered a weak pair. Moreover weak pairs are promoted to strong pairs when they are within the K closest neighbouring words calculated with a cosine distance. The skip-gram objective with negative sampling can be rephrased given the definition to positively and negatively couple words. The positive sampling cost function is

$$J_{pos}(w_t) = \beta_s \cdot \sum_{w_i \in V_s(w_t)} \ell(v_t \cdot v_i) \tag{22}$$

$$= \beta_w \cdot \sum_{w_j \in V_w(w_t)} \ell(v_t \cdot v_j) \tag{23}$$

ℓ is the logistic loss function, w_t is each target word of the corpus with its corresponding vector v_t, $V_s(w_t)$ are strong pairs, $V_w(w_t)$ are weak pairs and v_i/v_j are corresponding strong and weak pair vectors. The hyperparameters β_s and β_w are chosen to best fit to the learning of strong and weak pairs. When set to zero, the model behaves exactly like Word2Vec. The corresponding negative sampling cost function is

$$J_{neg}(w_t) = \sum_{w_i \in \mathcal{F}(w_t), w_i \notin \mathcal{S}(w_t), w_i \notin \mathcal{W}(w_t)} \ell(-v_t \cdot v_i) \tag{24}$$

Where w_i is chosen such that it is randomly chosen from the vocabulary at random without self $\mathcal{F}(w_t)$ and it is not part of strong $\mathcal{S}(w_t)$ or weak $\mathcal{W}(w_t)$ word pairs. Which results in the cost function J from a target word w_t with a context w_c: $J(w_t, w_c) = \ell(v_t \cdot v_c) + J_{pos}(w_t) + J_{neg}(w_t)$.

The results show an improvement over state-of-the-art models on word similarity and text classification. They parsed and trained on a last language corpus from Wikipedia comparing a pre-trained Word2Vec model augmented with dictionaries, a retrofitted model using WordNet and a single model on a raw corpus. Dict2Vec showed superior results on the raw corpus and improved the other models by up to 13%.

4.7   Context Selection

Context selection in language models is at this point a well studied tasks. Word2Vec uses a context window of surrounding words. While this sounds intuitive, there are a lot of suggestions on improving this. Originally, [22] suggested to use sub-sampling to remove frequently co-occurring words and use context distribution smoothing reducing bias towards rare words. This is very much in conjunction with count based methods that clip off the top/bottom percent of a vocabulary.

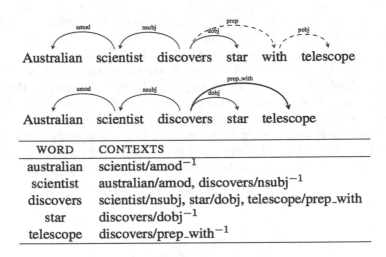

WORD	CONTEXTS
australian	scientist/amod^{-1}
scientist	australian/amod, discovers/nsubj^{-1}
discovers	scientist/nsubj, star/dobj, telescope/prep_with
star	discovers/dobj^{-1}
telescope	discovers/prep_with^{-1}

Fig. 2. Context capture as depicted in [19]

[19] have found that using dependency based word embeddings have an impact on the quality and quantity of functional similarity tasks such as *cosine*. However, it is to note that on topical similarity tasks the suggested model performs worse. [19] note that mostly a linear context, e.g., windows, is used. Given a corpora and a target word w, with a corresponding sentence (e.g. context) and modifiers of that sentence $m_1, \ldots, m_k$ with head h a dependency tree is created, see Fig. 2, with the Stanford Dependency parser.

The contexts $(m_1, l_1), \ldots, (m_k, l_k), (h, l_h^{-1})$, where l is the dependency relation between head and modifier (e.g. nsubj, dobj, prep with, amod). While l is the forward relation or outgoing relations from the head - the target word - l^{-1} is the in-going relation or inverse-relation. Given a Word2Vec model with a small window size of $k = 2$ and a larger window size $k = 5$ the dependency based model learns different word relations and minimizes two effects. We can see in Fig. 2 that coincidental filtering takes place, because "australian" is obviously not part of "science" in general, which Word2Vec would take as a context in either model. Secondly, if the window size is small out-of-reach words like "discover" and "telescope" would have been filtered out. Longer more complex sentences could have several head words where the context is out-of-reach in larger Word2Vec models as well. In comparison with Word2Vec the dependency base model has a higher precision and recall on functional similarity tests.

5 New Directions

Several new effective approaches in the area of language processing emerged in last two years. One of them bases on idea of attention-based modelling [35]. In contrast to previous approaches, it replaces complex recurrent or convolutional neural networks with a network architecture based solely on attention mechanisms. The authors in [35] call their application - the *Transformer*. The model architecture of the Transformer is shown in the Fig. 3.

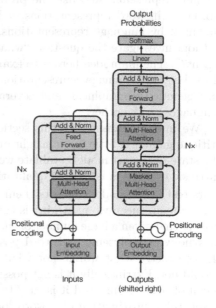

Authors of [31] present in their work an enhancement of the above approach by a generative pre-training of a language model. The pre-training proceeds completely unsupervised on a diverse corpus with long stretches of contiguous text. The authors present an improvement of the state of the art on 9 of the 12 datasets.

The application represents a semi-supervised approach as a combination of unsupervised pre-training and supervised

Fig. 3. The Transformer - model architecture as depicted in [35]

fine-tuning. In this way a universal representation can be learned and later transferred to a wide range of tasks. For the training of the universal representation large corpus of unlabeled text is used. The adaptation to specific tasks can be performed with much smaller pre-labeled data sets. This not only reduces the computation time but also increases the accuracy of the final specialized models.

A new type of deep contextualized word representation is introduced in [29]. They represent the word vectors (embeddings) as learned functions of the internal states of a deep bidirectional language model (biLM). This model is pre-trained on a large text corpus and can be added to existing models, improving the performance of different state-of-the-art NLP tasks. Such as question answering, textual entailment and sentiment analysis.

Another approach, see [1], introduces a new type of word embeddings - *contextual string embeddings*. The proposed embeddings are trained without explicit notion of words contextualized by their surrounding text.

According to authors, the same word will have different embeddings depending on its contextual use. They show in the experiments that this approach is especially useful for several downstream tasks, in particular for English and German named entity recognition.

6   Discussion

Currently is a time producing a lot of different models based on experimentation and educated guesses. It is usually left to the reader trying to find explanations in embeddings for language. What does a word-level embedding like Word2Vec actually represent? Just like the problem to separate semantic from syntactic similarity in word representations, it is not obvious what type of similarity is captured by language representations. In a new very interesting work [3] the authors investigate the question "what do language representations really represent?". The main idea here is to examine the correlations and causal relationships between language representations learned from translations on one hand, and genetic, geographical, and several levels of structural similarity between languages on the other.

We intend to discuss in this section some of the problems, challenges and critique gaining a little more insight on why embeddings actually work. Most of the state-of-the-art models evaluate word embeddings with intrinsic evaluations. Intrinsic evaluation is usually qualitative, given a set of semantic word analogy pairs test if the model connects them correctly: $\overrightarrow{man} - \overrightarrow{woman} \approx \overrightarrow{king} - \overrightarrow{queen}$. The woman/queen vs. man/king is the most famous of all examples. One could deduct that given a large number of such analogy word pairs, testing the presence of synonymy, polysemy and word positioning is sufficient. Intrinsic evaluation shows exactly what works, not what does not work or even what works but should not. Extrinsically it is not possible to use labels testing the precision and recall of our system. And it is easy to see why: What would you expect should a general approximation of a word look like? Should it be able to learn every possible dimension and therefore interpretation of what we perceive of it? If so, how should it learn to distinguish different domains with a different context? The context of a domain is never explained or given to the models.

Given a reasonable amount of test cases, quality can be ensured to some extent. How good or bad they actually perform is usually tested in downstream language tasks. If the embeddings perform better on that specific task compared to a preceding model, it is declared state-of-the-art. Interestingly, [6] shows that even state of the art embeddings display a large amount of bias towards certain topics: $\overrightarrow{man} - \overrightarrow{woman} \approx \overrightarrow{programmer} - \overrightarrow{homemaker}$.

Training real language models on real data yields real bias. The world and its written words are not fair and they incorporate really narrow views and concepts. Gender inequality and racism are two of the most challenging societal problems in the 21st century. Learning embeddings always yields a representation of the input. The bias is statistically significant. The problem is more obvious when considering that the standard Word2Vec model trained on the Googles News Corpus is applied on thousands of downstream language tasks. These kind of biases are not unique to language modelling and can be found in computer vision as well.

The authors in [6] hint that there are a three forms of bias: occupational stereotypes, analogies with stereotypes and indirect gender bias. They also acknowledge that not everything we perceive as bias should be seen as such e.g. *football* and *footballer* is male dominant for other reasons than just bias. To debias embeddings the answer is quite clear: we need additional knowledge in form of gender specific word lists. [6] suggest to create a reference model g with word vectors that are gender biased words.

While this works for a direct bias, it is much harder with indirect bias spread across different latent dimensions. Therefore a debiasing algorithm is suggested with two steps (1) Identify the gender subspace and (2) Equalize (factor out gender) or soften (reduce magnitude).

So far we have compared different word embedding models without looking into the theory. Intrinsic evaluations of new models yield better results and therefore improved models to be used. It seems however that there are no insights of actually why word embeddings are better than other models except for the experimentation. This seems plausible, as the de facto goal is to use an optimization procedure that has no closed mathematical form. The authors of [18] have found that Word2Vec with skip-gram and negative sampling is a PMI matrix. A (P)PMI matrix (extra P for keeping only positive entries) is a high dimensional and sparse context matrix where each row is a word w from the vocabulary V and each column represents a context c where it occurs. PPMI matrices are theoretically well known and provide a guiding hand for what Word2Vec actually learns. The problem of PPMI matrices is actually that you need to carefully consider each context for each occurring word, which does not scale up to billions of tokens. The results actually show that Word2Vec skip-gram with negative sampling is still the better choice from a view of precision and scalability. For further exploration of the theoretical aspects of word embeddings see [11] for an explanation of the additivity of vectors and [23] for a geometric interpretation of Word2Vec skip-gram with negative sampling.

7   Conclusion

In this paper we explored a wide variety of concepts dealing with word-level and subword-level embeddings as well as context selection procedures. This work represents an extension of our previous work in [27]. All of the suggested methods have assets and drawbacks. Besides what is covered here there are multiple research directions open. E.g., statistical models that treat words as a distribution, see [2,36]. [26] goes even further by representing words as hierarchical probability mass functions (pmfs). Instead of changing how the representation is created, they alter the representation to fit certain conditions and features. Most interesting approaches for the further research are those based on attention-based modelling [35] and unsupervised pre-training [31]. Due to significant improvements in computational efficiency and model performance, these approaches will dominate the future research in the area of Natural Language Processing.

References

1. Akbik, A., Blythe, D., Vollgraf, R.: Contextual string embeddings for sequence labeling. In: Proceedings of the 27th International Conference on Computational Linguistics, pp. 1638–1649 (2018)
2. Athiwaratkun, B., Wilson, A.: Multimodal word distributions. In: Conference of the Association for Computational Linguistics (ACL) (2017)
3. Bjerva, J., Östling, R., Han Veiga, M., Tiedemann, J., Augenstein, I.: What do language representations really represent? Comput. Linguist. 1–8 (2019, Just Accepted)
4. Blei, D., Ng, A., Jordan, M.: Latent dirichlet allocation. J. Mach. Learn. Res
5. Bojanowski, P., Grave, E., Joulin, A., Mikolov, T.: Enriching word vectors with subword information. CoRR
6. Bolukbasi, T., Chang, K., Zou, J., Saligrama, V., Kalai, A.: Man is to computer programmer as woman is to homemaker? Debiasing word embeddings. CoRR
7. Council, N.R., Committee, A.L.P.A.: Language and machines: computers in translation and linguistics, a report. In: National Academy of Sciences, National Research Council (1966)
8. Deerwester, S., Dumais, S., Furnas, G., Landauer, T., Harshman, R.: Indexing by latent semantic analysis. J. Am. Soc. Inform. Sci. **41**(6), 391–407 (1990)
9. Dhingra, B., Liu, H., Salakhutdinov, R., Cohen, W.: A comparative study of word embeddings for reading comprehension. CoRR
10. Dyer, C.: Notes on noise contrastive estimation and negative sampling. CoRR
11. Gittens, A., Achlioptas, D., Mahoney, M.: Skip-gram - zipf + uniform = vector additivity. In: Proceedings of the 55th Annual Meeting of the Association for Computational Linguistics, ACL 2017, Canada, vol. 1. pp. 69–76 (2017)
12. Graves, A., Schmidhuber, J.: Framewise phoneme classification with bidirectional LSTM and other neural network architectures. Neural Netw. 5–6 (2005)
13. Herbelot, A., Baroni, M.: High-risk learning: acquiring new word vectors from tiny data. CoRR
14. Joulin, A., Grave, E., Bojanowski, P., Mikolov, T.: Bag of tricks for efficient text classification. CoRR
15. Józefowicz, R., Vinyals, O., Schuster, M., Shazeer, N., Wu, Y.: Exploring the limits of language modeling. CoRR
16. Kim, Y., Jernite, Y., Sontag, D., Rush, A.: Character-aware neural language models. CoRR
17. Kneser, R., Ney, H.: Improved clustering techniques for class-based statistical language modelling. In: Third European Conference on Speech Communication and Technology (1993)
18. Levy, O., Goldberg, Y.: Neural word embedding as implicit matrix factorization. In: Advances in Neural Information Processing Systems, vol. 27
19. Levy, O., Goldberg, Y.: Dependency-based word embeddings. In: Proceedings of the 52nd Annual Meeting of the Association for Computational Linguistics, Baltimore, USA, vol. 2. pp. 302–308 (2014). http://aclweb.org/anthology/P/P14/P14-2050.pdf
20. Manning, C.D., Schuetze, H.: Foundations of Statistical Natural Language Processing. MIT Press, Cambridge (1999)
21. McCann, B., Bradbury, J., Xiong, C., Socher, R.: Learned in translation: contextualized word vectors. CoRR

22. Mikolov, T., Sutskever, I., Chen, K., Corrado, G., Dean, J.: Distributed representations of words and phrases and their compositionality. CoRR abs/1310.4546 (2013)
23. Mimno, D., Thompson, L.: The strange geometry of skip-gram with negative sampling. In: Proceedings of the 2017 Conference on Empirical Methods in Natural Language Processing, September 2017
24. Mnih, A., Hinton, G.: Three new graphical models for statistical language modelling. In: Proceedings of the 24th International Conference on Machine Learning, pp. 641–648. ACM (2007)
25. Ney, H., Essen, U., Kneser, R.: On structuring probabilistic dependences in stochastic language modelling. Comput. Speech Lang. **8**(1), 1–38 (1994)
26. Nickel, M., Kiela, D.: Poincaré embeddings for learning hierarchical representations. CoRR
27. Nitsche, M., Tropmann-Frick, M.: Context and embeddings in language modelling - an exploration. In: Selected Papers of the XX International Conference on Data Analytics and Management in Data Intensive Domains (DAMDID/RCDL 2018), Moscow, Russia, 9–12 October 2018, pp. 131–138 (2018). http://ceur-ws.org/Vol-2277/paper24.pdf
28. Pennington, J., Socher, R., Manning, C.: Glove: global vectors for word representation. In: Empirical Methods in Natural Language Processing (EMNLP), pp. 1532–1543 (2014). http://www.aclweb.org/anthology/D14-1162
29. Peters, M.E., et al.: Deep contextualized word representations. CoRR abs/1802.05365 (2018). http://arxiv.org/abs/1802.05365
30. Pinter, Y., Guthrie, R., Eisenstein, J.: Mimicking word embeddings using subword RNNs. CoRR
31. Radford, A., Narasimhan, K., Salimans, T., Sutskever, I.: Improving language understanding by generative pre-training (2018). https://s3-us-west-2.amazonaws.com/openai-assets/research-covers/languageunsupervised/languageunderstandingpaper.pdf
32. Rong, X.: word2vec parameter learning explained. CoRR abs/1411.2738 (2014). http://arxiv.org/abs/1411.2738
33. Srivastava, R., Greff, K., Schmidhuber, J.: Highway networks. CoRR
34. Tissier, J., Gravier, C., Habrard, A.: Dict2vec: learning word embeddings using lexical dictionaries. In: Proceedings of the Conference on Empirical Methods in Natural Language Processing, Copenhagen, Denmark, 9–11 September 2017, pp. 254–263 (2017)
35. Vaswani, A., et al.: Attention is all you need. CoRR abs/1706.03762 (2017). http://arxiv.org/abs/1706.03762
36. Vilnis, L., McCallum, A.: Word representations via Gaussian embedding. CoRR
37. Wieting, J., Bansal, M., Gimpel, K., Livescu, K.: Charagram: Embedding words and sentences via character n-grams. CoRR
38. Winograd, T.: Understanding natural language. Cogn. Psychol. **3**(1), 1–191 (1972)

Distributed Computing

Scalable Algorithm for Subsequence Similarity Search in Very Large Time Series Data on Cluster of Phi KNL

Yana Kraeva and Mikhail Zymbler[(✉)] [ID]

South Ural State University, Chelyabinsk, Russia
{kraevaya, mzym}@susu.ru

Abstract. Nowadays, subsequence similarity search under the Dynamic Time Warping (DTW) similarity measure is applied in a wide range of time series mining applications. Since the DTW measure has a quadratic computational complexity w.r.t. the length of query subsequence, a number of parallel algorithms for various many-core architectures have been developed, namely FPGA, GPU, and Intel MIC. In this paper, we propose a novel parallel algorithm for subsequence similarity search in very large time series data on computing cluster with nodes based on the Intel Xeon Phi Knights Landing (KNL) many-core processors. Computations are parallelized both at the level of all cluster nodes through MPI, and within a single cluster node through OpenMP. The algorithm involves additional data structures and redundant computations, which make it possible to effectively use Phi KNL for vector computations. Experimental evaluation of the algorithm on real-world and synthetic datasets shows that it is highly scalable.

Keywords: Time series · Similarity search · Dynamic Time Warping · Parallel algorithm · Cluster · OpenMP · MPI · Intel Xeon Phi · Knights Landing · Data layout · Vectorization

1 Introduction

Nowadays, time series are pervasive in a wide spectrum of applications with data intensive analytics, e.g. climate modelling [1], economic forecasting [21], medical monitoring [6], etc. Many time series analytical problems require subsequence similarity search as a subtask, which assumes the following. A query subsequence and a longer time series are given, and a subsequence of the time series should be found, whose similarity to the query is the maximum among all the subsequences.

Currently, Dynamic Time Warping (DTW) [3] is considered as the best similarity measure in most domains [5]. Since computation of DTW is time-consuming there are parallel algorithms for FPGA [25] and GPU [17] have been proposed.

Our research [10–13] addresses the task of accelerating similarity search with the Intel Xeon Phi many-core system, which can be considered as an attractive alternative to FPGA and GPU. Phi provides a large number of compute cores with 512-bit wide vector processing units. Phi is based on the Intel x86 architecture and supports the same

© Springer Nature Switzerland AG 2019
Y. Manolopoulos and S. Stupnikov (Eds.): DAMDID/RCDL 2018, CCIS 1003, pp. 149–164, 2019.
https://doi.org/10.1007/978-3-030-23584-0_9

programming methods and tools as a regular Intel Xeon. The first generation of Phi, Knights Corner (KNC) [4], is a coprocessor with up to 61 cores, which supports native applications and offloading of calculations from a host CPU. The second generation product, Knights Landing (KNL) [20], is a bootable processor with up to 72 cores, which runs applications only in native mode. In [11–13], we proposed CPU+Phi computational scheme for subsequence similarity search on Phi KNC. In [10], we changed such an approach for Phi KNL having implemented advanced data layout and computational scheme, which allow to efficiently vectorizing computations.

This paper is a revised extended version of [10]. We consider more complicated case of very large time series when computing cluster system of Phi KNL nodes is utilized for the similarity search. We propose an advanced parallel algorithm, called *PhiBestMatch*, which parallelizes computations both among cluster nodes (through MPI technology), and within a single cluster node (through OpenMP technology). We performed additional series of experiments, which showed good scalability of *PhiBestMatch*.

The rest of the paper is organized as follows. Section 2 discusses related work. Section 3 gives formal statement of the problem. In Sect. 4, we present the proposed algorithm. We describe experimental evaluation of our algorithm in Sect. 5. Finally, Sect. 6 concludes the paper.

2   Related Work

In recent decade, parallel and distributed algorithms for subsequence similarity search under the DTW measure have been extensively developed for various hardware platforms.

In [26], a GPU-based implementation was proposed. The warping matrix is generated in parallel, but the warping path is searched serially. Since the matrix generation step and the path search step are split into two kernels, this leads to overheads for storage and transmission of the warping matrix for each DTW calculation.

In [17], GPU and FPGA implementations of subsequence similarity search were presented. The GPU implementation is based on the same ideas as [26]. The system consists of two modules, namely Normalizer (z-normalization of subsequences) and Warper (DTW calculation), and is generated by a C-to-VHDL tool, which exploits the fine-grained parallelism of the DTW. However, this implementation suffers from lacking flexibility, i.e. it must be recompiled if length of query is changed. In [25], authors proposed a framework for FPGA-based subsequence similarity search, which utilizes the data reusability of continuous DTW calculations to reduce the bandwidth and exploit the coarse-grain parallelism.

In [22], authors proposed subsequence similarity search on CPU cluster. Subsequences starting from different positions of the time series are sent to different nodes, and each node calculates DTW in the naïve way. In [23], authors accelerated subsequence similarity search with SMP system. They distribute different queries into different cores, and each subsequence is sent to different cores to be compared with different patterns in the naïve way. In both implementations, the data transfer becomes the bottleneck. In [18], authors proposed an approach to subsequence similarity search on Apache Spark cluster. Time series is fragmented and fragments are shared among cluster nodes as files under HDFS (Hadoop Distributed File System). Each node processes in parallel as many fragments as its CPU cores where each core implements the UCR-DTW algorithm [15].

3 Notation and Problem Background

3.1 Definitions and Notation

A *time series* T is a sequence of real-valued elements: $T = (t_1, t_2, \ldots, t_m)$. Length of a time series T is denoted by $|T|$.

Given two time series, $X = (x_1, x_2, \ldots, x_m)$ and $Y = (y_1, y_2, \ldots, y_m)$, the *Dynamic Time Warping (DTW)* distance between X and Y is denoted by $DTW(X, Y)$ and defined as below.

$$DTW(X, Y) = d(m, m), d(i, j) = (x_i - y_j)^2 + \min \begin{cases} d(i-1, j) \\ d(i, j-1) \\ d(i-1, j-1) \end{cases}, \qquad (1)$$

$$d(0, 0) = 0, d(i, 0) = d(0, j) = \infty, 1 \le i \le m, 1 \le j \le m.$$

In the formulas above, $(d_{ij}) \in \mathbb{R}^{m \times m}$ is considered as a *warping matrix* for the alignment of the two respective time series. A *warping path* is a contiguous set of warping matrix elements that defines a mapping between two time series. The warping path must start and finish in diagonally opposite corner cells of the warping matrix, the steps in the warping path are restricted to adjacent cells, and the points in the warping path must be monotonically spaced in time.

A *subsequence* $T_{i,k}$ of a time series T is its contiguous subset of k elements, which starts from position i: $T_{i,k} = (t_i, t_{i+1}, \ldots, t_{i+k-1}), 1 \le i \le m - k + 1$. A set of all subsequences of T with length n is denoted by S_T^n. Let $N = |T| - n + 1 = m - n + 1$ denotes a number of subsequences in S_T^n.

Given a time series T and a time series Q as a user specified query where $m = |T| \gg |Q| = n$, the *best matching subsequence* $T_{i,n}$ meets the property

$$\exists T_{i,n} \in S_T^n \ \forall k \, DTW(Q, T_{i,n}) \le DTW(Q, T_{k,n}), 1 \le i, k \le m - n + 1. \qquad (2)$$

In what follows, where there is no ambiguity, we refer to subsequence $T_{i,n}$ as C, as a candidate in match to a query Q.

3.2 The UCR-DTW Serial Algorithm

Currently, UCR-DTW [15] is the fastest serial algorithm of subsequence similarity search, which integrates a large number of algorithmic speedup techniques. Since our algorithm is based on UCR-DTW, we briefly describe its basic features.

Squared Distances. The *Euclidean distance (ED)* between two subsequences Q and C where $|Q| = |C|$, is defined as below.

$$ED(Q, C) = \sqrt{\sum_{i=1}^{n} (q_i - c_i)^2}. \tag{3}$$

Instead of use square root in DTW and ED distance calculation, it is possible to use the squares thereof since it does not change the relative rankings of subsequences.

Z-normalization. Both the query subsequence and each subsequence of the time series need to be z-normalized before the comparison [24]. The *z-normalization* of a time series T is defined as a time series $\hat{T} = (\hat{t}_1, \hat{t}_2, \ldots, \hat{t}_m)$ where

$$\hat{t}_i = \frac{t_i - \mu}{\sigma}, \quad \mu = \frac{1}{m} \sum_{i=1}^{m} t_i, \quad \sigma^2 = \frac{1}{m} \sum_{i=1}^{2} t_i^2 - \mu^2. \tag{4}$$

Cascading Lower Bounds. Lower bound (LB) is an easy computable threshold of the DTW distance measure to identify and prune clearly dissimilar subsequences [5]. In what follows, we refer this threshold as the *best-so-far* distance (or *bsf* for brevity). If LB has exceeded *bsf*, the DTW distance will exceed *bsf* as well, and the respective subsequence is assumed to be clearly dissimilar and pruned without calculation of DTW. UCR-DTW initializes *bsf* as $+\infty$ and then scans the time series with sliding window and calculates *bsf* on the *i*th step as follows:

$$bsf_{(i)} = \min \left(bsf_{(i-1)}, \begin{cases} +\infty, LB(Q, T_{i,n}) > bsf_{(i-1)} \\ DTW(Q, T_{i,n}), otherwise \end{cases} \right). \tag{5}$$

UCR-DTW exploits three LBs, namely $LB_{Kim}FL$ [15], $LB_{Keogh}EC$, $LB_{Keogh}EQ$ [8] applying them in a cascade.

The $LB_{Kim}FL$ lower bound uses the distances between the First (Last) pair of points from C and Q as a lower bound, and defined as below.

$$LB_{Kim}FL(Q, C) := ED(\hat{q}_1, \hat{c}_1) + ED(\hat{q}_n, \hat{c}_n) \tag{6}$$

The $LB_{Keogh}EC$ lower bound is the distance from the closer of the two so-called envelopes of the query to a candidate subsequence, and defined as below.

$$LB_{Keogh}EC(Q, C) = \sum_{i=1}^{n} \begin{cases} (\hat{c}_i - u_i)^2, if \ \hat{c}_i > u_i \\ (\hat{c}_i - \ell_i)^2, if \ \hat{c}_i < \ell_i \\ 0, otherwise \end{cases} \tag{7}$$

In the equation above, subsequences $U = (u_1, \ldots, u_n)$ and $L = (\ell_1, \ldots, \ell_n)$ are the *upper envelope* and *lower envelope* of the query, respectively, and defined as below.

$$u_i = \max_{i-r \leq k \leq i+r} \hat{q}_k, \ell_i = \min_{i-r \leq k \leq i+r} \hat{q}_k, \tag{8}$$

where the parameter r $(1 \leq r \leq n)$ denotes the Sakoe–Chiba band constraint [16], which states that the warping path cannot deviate more than r cells from the diagonal of the warping matrix.

The $LB_{Keogh}EQ$ lower bound is the distance from the query and the closer of the two envelopes of a candidate subsequence (i.e. the roles of the query and the candidate subsequence are reversed as opposed to $LB_{Keogh}EC$).

$$LB_{Keogh}EQ(Q, C) := LB_{Keogh}EC(C, Q). \tag{9}$$

Firstly, UCR-DTW calculates z-normalized version of the query and its envelopes, and bsf is assumed to be equal to infinity. Then the algorithm scans the input time series applying the cascade of LBs to the current subsequence. If the subsequence is not pruned, then DTW distance is calculated. Next, bsf is updated if it is greater than the value of DTW distance calculated above. By doing so, in the end, UCR-DTW finds the best matching subsequence of the given time series.

4 The *PhiBestMatch* Parallel Algorithm

In this section, we present a novel parallel algorithm for subsequence similarity search in very long time series on computing cluster of Phi KNL nodes, called *PhiBestMatch*. *PhiBestMatch* is based on the following ideas.

Computations are parallelized on two levels, namely at the level of all cluster nodes, and within a single cluster node. The time series is divided into equal-length partitions and distributed among cluster nodes. During the search in its own partition, each node communicates with rest nodes by functions of the MPI standard to improve local bsf and reduce the amount of computations.

Within a single cluster node, computations are performed by the thread-level parallelism and the OpenMP technology. In addition, data structures are aligned in main memory, and computations are organized with as many vectorizable loops as possible. Vectorization means a compiler's ability to transform the loops into sequences of vector operations [2] of VPUs. We should avoid unaligned memory access since it can cause inefficient vectorization due to timing overhead for loop peeling [2]. Within a single cluster node, the algorithm involves additional data structures and redundant computations [10].

4.1 Partitioning of the Time Series

We partition the time series among cluster nodes as follows. Let F is a number of fragments and $T^{(k)}$ is k-th $(0 \leq k \leq F - 1)$ partition of T, then $T^{(k)}$ is defined as a subsequence $T_{start,len}$ as below.

$$start = k \cdot \left\lfloor \frac{N}{F} \right\rfloor + 1, \; len = \begin{cases} \left\lfloor \frac{N}{F} \right\rfloor + (N \bmod F) + n - 1, k = F - 1 \\ \left\lfloor \frac{N}{F} \right\rfloor + n - 1, otherwise \end{cases}. \quad (10)$$

This means the head part of every partition except first overlaps with the tail part of the previous partition in $n - 1$ data points, where n is the query length. Such a technique prevents us from loss of the resulting subsequences in the junctions of two neighbor partitions.

4.2 Data Layout

We propose data layout aiming to provide organize computations over aligned data with as many auto-vectorizable loops as possible.

Given a subsequence C and VPU width w, we denote *pad length* as $pad = w - (n \bmod w)$ and define *aligned subsequence* $\tilde{T}_{i,n}$ as below:

$$\tilde{T}_{i,n} = \begin{cases} (t_i, t_{i+1}, \ldots, t_{i+n-1}, \underbrace{0, 0, \ldots, 0}_{pad}), & if \; n \bmod w > 0 \\ \\ (t_i, t_{i+1}, \ldots, t_{i+n-1}), & otherwise. \end{cases} \quad (11)$$

According to (1), $\forall Q, C : |Q| = |C| \; DTW(Q, C) = DTW(\tilde{Q}, \tilde{C})$. Thus, in what follows, we will assume the aligned versions of the query and a subsequence of the input time series.

Next, we store all (aligned) subsequences of a time series in the *subsequence matrix* $S_T^n \in \mathbb{R}^{N \times (n + pad)}$, which is defined as below.

$$S_T^n(i, j) := \tilde{t}_{i+j-1}. \quad (12)$$

Let us denote the number of LBs exploited by the algorithm as lb_{max} ($lb_{max} \geq 1$), and denote these LBs as $LB_1, LB_2, \ldots, LB_{lb_{max}}$, enumerating them according to the order in the lower bounding cascade. Given a time series T, we define the *LB-matrix* of all subsequences of length n from T, $L_T^n \in \mathbb{R}^{N \times lb_{max}}$ as below.

$$L_T^n(i, j) := LB_j(T_{i,n}, Q). \quad (13)$$

The *bitmap matrix* is a vector-column $B_T^n \in \mathbb{B}^N$, which for all subsequences of length n from T stores the logical conjunction of *bsf* and every LB:

$$B_T^n(i) := \bigwedge_{j=1}^{lb_{max}} \left(L_T^n(i, j) < bsf \right). \quad (14)$$

We establish the *candidate matrix* to store those subsequences from the S_T^n matrix, which have not been pruned after the lower bounding. The candidate matrix will be processed in parallel by calculating of DTW distance measure between each row of the matrix and the query. Then the minimum of DTW distances is used as *bsf*.

To provide parallel calculations of the candidate matrix, we denote the *segment size of the matrix as* $s \in \mathbb{N}$ ($s \leq \frac{N}{p}$ where p is the number of threads employed by the parallel algorithm) and define the *candidate matrix*, $C_T^n \in \mathbb{R}^{(s \cdot p) \times (n + pad)}$ as below.

$$C_T^n(i, \cdot) := S_T^n(k, \cdot) \colon B_T^n(i) = TRUE. \tag{15}$$

In further experiments, we take the segment size $s = 100$.

4.3   Computational Scheme

Figure 1 depicts the *PhiBestMatch* pseudo-code, and Fig. 2 shows data structures of the algorithm. At initialization, the algorithm assigns the number of the current process to *myrank* by the MPI function. In what follows, each process deals with the subsequence matrix $S_{T^{(myrank)}}^n$ of the $T^{(myrank)}$ partition. The variable *bsf* is initialized by the DTW distance between the query and a random subsequence of the partition.

Algorithm PHIBESTMATCH
  Input:
    T time series to search
    Q query subsequence
    r warping constraint
  Output:
    bsf similarity of the best match subsequence
    bestmatch   index of the best match subsequence

1: $myrank \leftarrow$ MPI_Comm_rank()
2: $N \leftarrow |N^{(myrank)}| - n + 1$
3: $subseq_{rnd} \leftarrow T_{random(1..N),n}^{(myrank)}$
4: $bsf \leftarrow$ DTW($subseq_{rnd}, Q, r, \infty$)
5: $myrank \leftarrow N$
6: PREPROCESS($T^{(myrank)}, Q, S_T^n, L_T^n$)
7: **repeat**
8:      IMPROVE($T^{(myrank)}, bsf, bestmatch$)
9:      $flagDone \leftarrow (processed = 0)$
10:    { $bsf, bestmatch$ } $\leftarrow$ MPI_Allreduce({$bsf, bestmatch$}, MPI_FLOAT_LONG, MPI_MIN)
11:    $Stop \leftarrow$ MPI_Allreduce($myFlagDone$, MPI_BOOL, MPI_AND)
12: **until not** *Stop*
13: **return** {$bsf, bestmatch$}

Fig. 1. *PhiBestMatch* pseudo-code

Then we perform preprocessing by forming the subsequence matrix of the aligned subsequences, z-normalizing each subsequence, and calculating each LB of the lower bounding cascade. Strictly speaking, the latter step brings redundant calculations. In contrast, UCR-DTW calculates the next LB in the cascade only if a current subsequence is not clearly dissimilar after the calculation of the previous LB. However, we perform precomputations once and parallelize them keeping in mind they further can be efficiently vectorized by the compiler since the absence of data dependencies in LBs.

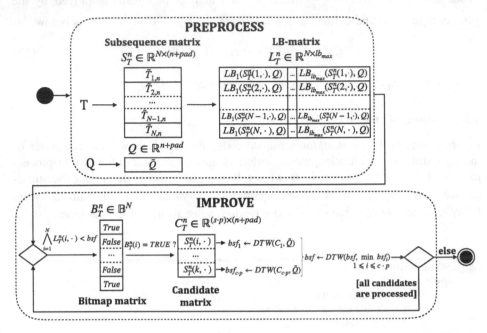

Fig. 2. Data flow of *PhiBestMatch*

After that, the algorithm improves the *bsf* threshold by the following loop until each node completes its partition. At first, the bitmap matrix is calculated in parallel based on the pre-calculated LB-matrix. Then each subsequence with TRUE in the respective element of the bitmap matrix is added to the candidate matrix. After the candidate matrix is filled, we calculate in parallel the DTW distance measure between each candidate and the query and find the minimum distance. If the minimum distance is less than *bsf* then *bsf* is updated. Then, we find the minimum value of *bsf* among all the partitions by the MPI_Allreduce global reduction operation. Finally, the latter operation is used to check if each node completes its partition.

5 Experiments

In order to evaluate the developed algorithm, we performed experiments on two platforms, namely a single cluster node and a whole cluster system.

5.1 Experimental Setup

Objectives. In the experiments on a single cluster node, we studied performance and scalability of the algorithm with respect to the r warping constraint and the n query length. In the experiments on the cluster system, we studied the algorithm's scaled speedup with respect to the query length. Finally, we compare *PhiBestMatch* performance with analogous algorithm [18].

Measures. In the experiments, we investigated the algorithm's performance (measuring the run time after deduction of the I/O time) and scalability. We calculated the algorithm's speedup and parallel efficiency, which are defined as follows. *Speedup* and *parallel efficiency* of a parallel algorithm employing k threads are calculated, respectively, as

$$s(k) = \frac{t_1}{t_k}, e(k) = \frac{s(k)}{k}, \qquad (16)$$

where t_1 and t_k are run times of the algorithm when one and k threads are employed, respectively.

In the experiments on the cluster system, we investigated *scaled speedup* of the parallel algorithm, which refers to linear increasing of the problem size proportionally with the number of computational nodes added to the system, and is calculated as follows:

$$S_{scaled} = \frac{p \cdot m}{t_{p(p \cdot m)}}, \qquad (17)$$

where p is the number of nodes, m is the problem size, and $t_{p(p \cdot m)}$ is the algorithm's run time when a problem of size $p \cdot m$ is processed on p nodes.

Hardware. We performed our experiments on two supercomputers, namely Tornado SUSU [9] and NKS-1P [19] with the characteristics summarized in Table 1.

Table 1. Specifications of hardware

Specifications	Tornado SUSU		NKS-1P	
	Host	Node	Host	Node
Model, Intel Xeon	2 × X5680	Phi KNC, SE10X	2 × E5-2630v4	Phi KNL 7290
Physical cores	2 × 6	61	2 × 10	72
Hyper threading factor	2	4	2	4
Logical cores	24	244	40	288
Frequency, GHz	3.33	1.1	2.2	1.5
VPU width, bit	128	512	256	512
Peak performance, TFLOPS	0.371	1.076	0.390	3.456

For the experiments on a single node, we used the simplified version of *PhiBestMatch* [10], which treats the time series as one partition.

Datasets. In the experiments, we used datasets summarized in Table 2. RW-SN, RW-CS, and RW-SN are the datasets generated according to the Random Walk model [14]. The EPG (Electrical Penetration Graph) dataset is a series of signals, which was used by entomologists to study of Aster leafhopper (*macrosteles quadrilineatus*) behavior [17]. The ECG dataset [7] represents electrocardiogram signals digitized at 128 Hz.

Table 2. Datasets used in experiments

Platform	Dataset	Type	$|T| = m$	$|Q| = n$
Single cluster node	RW-SN	Synthetic	10^6	128
Single cluster node	EPG	Real	$2.5 \cdot 10^5$	360, 432, 512, 1024
Cluster system	RW-CS	Synthetic	$12.8 \cdot 10^7$	128, 512, 1024
Cluster system	ECG	Real	$12.8 \cdot 10^7$	432, 512, 1024
Cluster system	RW-SH	Synthetic	$2.2 \cdot 10^8$	128

5.2   Evaluation on a Single Cluster Node

Figures 3 and 4 depict the performance of *PhiBestMatch* depending on r and n, respectively. As we can see, at lower values of the parameters (approximately, $0 < r \leq 0.5n$ and $n < 512$), the algorithm runs slightly faster or about the same way on two Intel Xeon host than on Intel Xeon Phi. At high values of the parameters ($0.5n < r \leq n$ and $n \geq 512$), the algorithm is faster on Intel Xeon Phi. It means that *PhiBestMatch* better utilizes vectorization capabilities of Intel Xeon Phi with greater computational load.

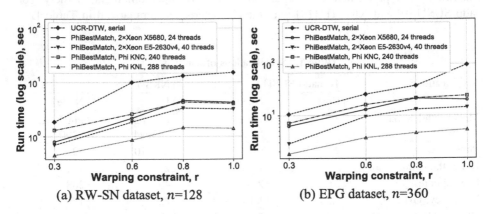

(a) RW-SN dataset, n=128   (b) EPG dataset, n=360

Fig. 3. *PhiBestMatch* performance w.r.t. the warping constraint

Figures 5 and 6 depict the experimental results on the synthetic (RW-SN) and the real (EPG) datasets, respectively. As we can see, *PhiBestMatch* shows speedup closer to linear and efficiency closer to 100%, if the number of threads matches the number of

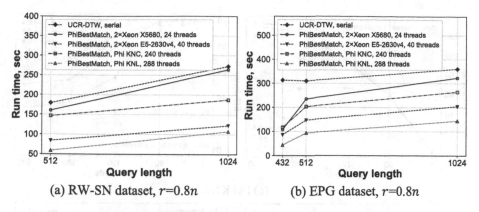

(a) RW-SN dataset, $r=0.8n$ (b) EPG dataset, $r=0.8n$

Fig. 4. *PhiBestMatch* performance w.r.t. the query length

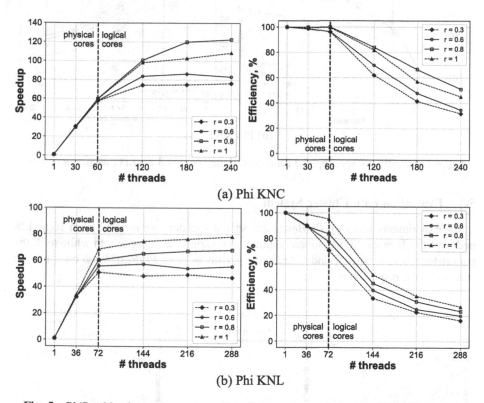

Fig. 5. *PhiBestMatch* speedup and parallel efficiency on synthetic data (RW-SN dataset)

physical cores the algorithm is running on. When more than one thread per physical core is used, speedup became sub-linear, and parallel efficiency decreases accordingly. The best speedup and efficiency are achieved when the r parameter ranges from 0.8 to 1 of n.

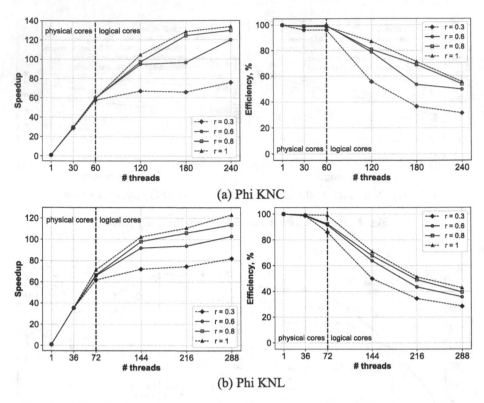

(a) Phi KNC

(b) Phi KNL

Fig. 6. *PhiBestMatch* speedup and parallel efficiency on real data (EPG dataset, $n = 360$)

5.3   Evaluation on a Cluster System

In the experiments studying *PhiBestMatch* scaled speedup, we utilized from 16 to 128 nodes of the Tornado SUSU supercomputer. We varied the query length while took the parameter $r = n$. Figures 7 and 8 depict the performance of *PhiBestMatch* on synthetic and real data, respectively.

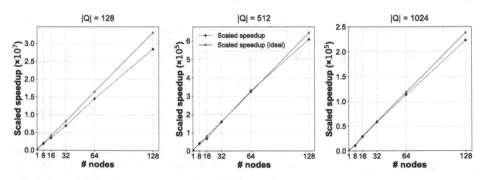

Fig. 7. *PhiBestMatch* scaled speedup on synthetic data (RW-CS dataset, $r = 0.8n$)

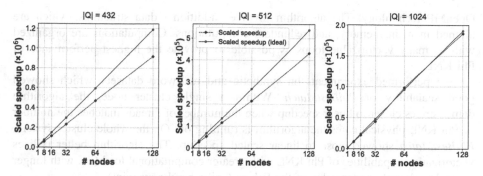

Fig. 8. *PhiBestMatch* scaled speedup on real data (ECG-CS dataset, $r = 0.8n$)

As we can see, *PhiBestMatch* shows closer to linear scaled speedup. At the same time, the similarity search for a subsequence of greater length demonstrates a higher scaled speedup, since it provides a greater amount of computations on a single node.

5.4 Comparison with Analogue

In [18], Shabib *et al.* presented the hybrid search algorithm, which exploits Apache Spark cluster of multi-core nodes. The algorithm was evaluated on six cluster nodes each with Intel Xeon E3-1200 (4-core at 3.1 GHz) CPU onboard for the RW-SH dataset with the parameter $r = 0.05n$. We compared the performance of Shabib *et al.* algorithm and *PhiBestMatch* performance on six nodes of Tornado SUSU for the same dataset and parameter r. Table 4 depicts the results.

Table 4. Performance of *PhiBestMatch* in comparison with Shabib *et al.* algorithm

PhiBestMatch, sec	Algorithm of Shabib et al., sec
24.2	32

6 Conclusion

In this paper, we presented *PhiBestMatch*, a novel parallel algorithm for subsequence similarity search in very large time series data on computing cluster of the modern Intel Xeon Phi Knights Landing (Phi KNL) nodes. Phi KNL is many-core system with 512-bit wide vector processing units, which supports the same programming methods and tools as a regular Intel Xeon, and can be considered as an alternative to FPGA and GPU.

PhiBestMatch performs parallel computations on two levels, namely at the level of all cluster nodes, and within a single cluster node. The time series is divided into equal-length partitions and distributed among cluster nodes. During the search in its own partition, each node communicates with rest nodes by functions of the MPI standard to improve local best-so-far similarity threshold and reduce the amount of computations. Within a single cluster node, *PhiBestMatch* exploits the thread-level parallelism and the

OpenMP technology. The algorithm involves additional data structures, which are aligned in main memory, and redundant computations. Computations are organized with as many vectorizable loops as possible to provide the highest performance of Phi KNL.

We performed experiments on synthetic and real-word datasets, which showed good scalability of *PhiBestMatch*. Within a single cluster node, the algorithm demonstrates closer to linear speedup when the number of threads matches the number of Phi KNL physical cores the algorithm is running on. On the whole cluster system, *PhiBestMatch* showed close to linear scaled speedup. The algorithm better utilizes vectorization capabilities of Phi KNL with greater computational load (i.e. with longer query length and greater value of the Sakoe–Chiba band constraint).

Acknowledgments. This work was financially supported by the Russian Foundation for Basic Research (grant No. 17-07-00463), by Act 211 Government of the Russian Federation (contract No. 02.A03.21.0011) and by the Ministry of education and science of Russian Federation (government order 2.7905.2017/8.9). Authors thank The Siberian Branch of the Russian Academy of Sciences (SB RAS) Siberian Supercomputer Center (Novosibirsk, Russia) for the provided computational resources.

References

1. Abdullaev, S.M., Zhelnin, A.A., Lenskaya, O.Y.: The structure of mesoscale convective systems in central Russia. Russ. Meteorol. Hydrol. **37**(1), 12–20 (2012)
2. Bacon, D.F., Graham, S.L., Sharp, O.J.: Compiler transformations for high-performance computing. ACM Comput. Surv. **26**(4), 345–420 (1994). https://doi.org/10.1145/197405.197406
3. Berndt, D.J., Clifford, J.: Using dynamic time warping to find patterns in time series. In: Proceedings of the 1994 AAAI Workshop on Knowledge Discovery in Databases, Seattle, Washington, July 1994, pp. 359–370. AAAI Press (1994)
4. Chrysos, G.: Intel Xeon Phi coprocessor (codename Knights Corner). In: 2012 IEEE Hot Chips 24th Symposium (HCS), Cupertino, CA, USA, 27–29 August 2012, pp. 1–31. IEEE (2012). https://doi.org/10.1109/hotchips.2012.7476487
5. Ding, H., Trajcevski, G., Scheuermann, P., Wang, X., Keogh, E.: Querying and mining of time series data: experimental comparison of representations and distance measures. Proc. VLDB Endow. **1**(2), 1542–1552 (2008). https://doi.org/10.14778/1454159.1454226
6. Epishev, V., Isaev, A., Miniakhmetov, R., et al.: Physiological data mining system for elite sports. Bull. South Ural State Univ. Ser. Comput. Math. Softw. Eng. **2**(1), 44–54 (2013)
7. Goldberger, A.L., Amaral, L.A.N., Glass, L., Hausdorff, J.M., Ivanov, Pl.Ch., et al.: PhysioBank, PhysioToolkit, and PhysioNet. Circulation **101**(23), e215–e220 (2000). https://doi.org/10.1161/01.cir.101.23.e215
8. Keogh, E.J., Ratanamahatana, C.A.: Exact indexing of dynamic time warping. Knowl. Inf. Syst. **7**(3), 358–386 (2005). https://doi.org/10.1007/s10115-004-0154-9
9. Kostenetskiy, P., Semenikhina, P.: SUSU supercomputer resources for industry and fundamental science. In: GloSIC 2018, Proceedings of the Global Smart Industry Conference, Chelyabinsk, Russia, 13–15 November 2018, Article no. 8570068 (2018). https://doi.org/10.1109/glosic.2018.8570155

10. Kraeva, Ya., Zymbler, M.: An efficient subsequence similarity search on modern Intel many-core processors for data intensive applications. In: Proceedings of the 20th International Conference on Data Analytics and Management in Data Intensive Domains (DAMDID/RCDL 2018). CEUR Workshop Proceedings, Moscow, Russia, 9–12 October 2018, vol. 2277, pp. 143–151. CEUR-WS.org (2018)
11. Movchan, A.V., Zymbler, M.L.: Parallel algorithm for local-best-match time series subsequence similarity search on the Intel MIC architecture. Procedia Comput. Sci. **66**, 63–72 (2015). https://doi.org/10.1016/j.procs.2015.11.009%5d
12. Movchan, A.V., Zymbler, M.L.: Parallel implementation of searching the most similar subsequence in time series for computer systems with distributed memory. In: Sokolinsky, L., Starodubov, I. (eds.) PCT 2016, International Scientific Conference on Parallel Computational Technologies. CEUR Workshop Proceedings, Arkhangelsk, Russia, 29–31 March 2016, vol. 1576, pp. 615–628. CEUR-WS.org (2016)
13. Movchan, A., Zymbler, M.: Time series subsequence similarity search under dynamic time warping distance on the intel many-core accelerators. In: Amato, G., Connor, R., Falchi, F., Gennaro, C. (eds.) SISAP 2015. LNCS, vol. 9371, pp. 295–306. Springer, Cham (2015). https://doi.org/10.1007/978-3-319-25087-8_28
14. Pearson, K.: The problem of the random walk. Nature **72**(1865), 294 (1905). https://doi.org/10.1038/072342a0
15. Rakthanmanon, T., Campana, B.J.L., Mueen, A., Batista, G.E.A.P.A., Westover, M.B., et al.: Searching and mining trillions of time series subsequences under dynamic time warping. In: Proceedings of the 18th ACM SIGKDD International Conference on Knowledge Discovery and Data Mining, Beijing, China, 12–16 August 2012, pp. 262–270. ACM, New York (2012). https://doi.org/10.1145/2339530.2339576
16. Sakoe, H., Chiba, S.: Dynamic Programming algorithm optimization for spoken word recognition. In: Waibel, A., Lee, K.-F. (eds.) Readings in Speech Recognition, pp. 159–165. Morgan Kaufmann Publishers Inc., San Francisco (1990)
17. Sart, D., Mueen, A., Najjar, W.A., Keogh, E.J., Niennattrakul, V.: Accelerating dynamic time warping subsequence search with GPUs and FPGAs. In: Proceedings of the 2010 IEEE International Conference on Data Mining, Sydney, Australia, 14–17 December 2010, pp. 1001–1006. IEEE Computer Society, Washington, DC (2010). https://doi.org/10.1109/icdm.2010.21
18. Shabib, A., Narang, A., Niddodi, C.P., et al.: Parallelization of searching and mining time series data using dynamic time warping. In: Proceedings of the 2015 International Conference on Advances in Computing, Communications and Informatics, Kochi, India, 10–13 August, 2015, pp. 343–348. IEEE (2015). https://doi.org/10.1109/icacci.2015.7275633
19. Siberian Supercomputing Centre of ICMMG SB RAS. http://www.sscc.icmmg.nsc.ru/hardware.html
20. Sodani, A.: Knights Landing (KNL): 2nd generation Intel Xeon Phi processor. In: 2015 IEEE Hot Chips 27th Symposium (HCS), Cupertino, CA, USA, 22–25 August 2015, pp. 1–24. IEEE (2015)
21. Sokolinskaya, I., Sokolinsky, L.: Revised pursuit algorithm for solving non-stationary linear programming problems on modern computing clusters with manycore accelerators. Commun. Comput. Inf. Sci. **687**, 212–223 (2016). https://doi.org/10.1007/978-3-319-55669-7_17
22. Srikanthan, S., Kumar, A., Gupta, R.: Implementing the dynamic time warping algorithm in multithreaded environments for real time and unsupervised pattern discovery. In: 2011 2nd International Conference on Computer and Communication Technology, Allahabad, India, 15–17 September 2011, pp. 394–398. IEEE (2015). https://doi.org/10.1109/iccct.2011.6075111

23. Takahashi, N., Yoshihisa, T., Sakurai, Y., Kanazawa, M.: A parallelized data stream processing system using dynamic time warping distance. In: 2009 International Conference on Complex, Intelligent and Software Intensive Systems, Fukuoka, Japan, 16–19 March 2009, pp. 1100–1105. IEEE (2009). https://doi.org/10.1109/cisis.2009.77
24. Tarango, J., Keogh, E.J., Brisk, P.: Instruction set extensions for dynamic time warping. In: Proceedings of the International Conference on Hardware/Software Codesign and System Synthesis, Montreal, QC, Canada, 29 September–4 October 2013, pp. 18:1–18:10. IEEE (2013). https://doi.org/10.1109/codes-isss.2013.6659005
25. Wang, Z., Huang, S., Wang, L., Li, H., Wang, Y., et al.: Accelerating subsequence similarity search based on dynamic time warping distance with FPGA. In: Proceedings of the ACM/SIGDA International Symposium on Field Programmable Gate Arrays, Monterey, CA, USA, 11–13 February 2013, pp. 53–62. ACM, New York (2013). https://doi.org/10.1145/2435264.2435277
26. Zhang, Y., Adl, K., Glass, J.R.: Fast spoken query detection using lower-bound dynamic time warping on graphical processing units. In: 2012 IEEE International Conference on Acoustics, Speech and Signal Processing, Kyoto, Japan, 25–30 March 2012, pp. 5173–5176. IEEE (2012). https://doi.org/10.1109/icassp.2012.6289085

Information Extraction from Text

Neural Network Approach
for Extracting Aggregated Opinions
from Analytical Articles

Nicolay Rusnachenko[1(✉)] and Natalia Loukachevitch[2(✉)]

[1] Bauman Moscow State Technical University, Moscow, Russia
kolyarus@yandex.ru
[2] Lomonosov Moscow State University, Moscow, Russia
louk_nat@mail.ru

Abstract. Large texts that analyze a situation in some domain, for example politics or economy, usually are full of opinions. In case of analytical articles, opinions usually are a kind of attitudes with source and target presented as named entities, both mentioned in the text. We present an application of the specific neural network model for sentiment attitude extraction. This problem is considered as a three-class machine learning task for the whole documents. Treating text attitudes as a list of related contexts, we first extract related sentiment contexts and then calculate the resulted attitude label. For sentiment context extraction, we use Piecewise Convolutional Neural Network (PCNN). We experiment with variety of functions that allows us to compose the attitude label, including recurrent neural network, which give the possibility to take into account additional context aspects. For experiments, the RuSentRel corpus was used, it contains Russian analytical texts in the domain of international relations.

Keywords: Sentiment analysis · Convolutional Neural Networks · Relation extraction

1 Introduction

Automatic sentiment analysis, i.e. the identification of the author's opinion on the subject discussed in the text, is one of the most popular applications of natural language processing, during the last years. Amount of occured opinions and their sources in text significantly varies and depends on document genres.

Users' reviews towards services or products represent one of the most popular document genres in sentiment analysis tasks. In case of opinions from microblogging social networks [9], we deal with limited in length texts, which usually discuss a single entity (but, perhaps in its various aspects), and the opinion is expressed by the author of the message [1,12].

This work is partially supported by RFBR grant N 16-29-09606.

© Springer Nature Switzerland AG 2019
Y. Manolopoulos and S. Stupnikov (Eds.): DAMDID/RCDL 2018, CCIS 1003, pp. 167–179, 2019.
https://doi.org/10.1007/978-3-030-23584-0_10

Analytical articles represent another genre of documents for sentiment analysis, and differ from reviews with complicated discorse structure. These texts contain opinions conveyed by different subjects, including the author(s)' attitudes, positions of cited sources, and relations of the mentioned entities between each other. Texts usually mention a lot of named entities, and only a few of them are sources or targets of sentiment attitudes.

We describe a problem of sentiment attitude extraction from analytical articles written in Russian. Considering *attitudes* as directed relations between mentioned named entities, the task is to extract only sentiment of them. This paper proceeds the work [13] by presenting proposed approach in more details and extend it with application of various label aggregation functions.

2   Related Work

The task of attitude recognition toward named entities or events including opinion holder identification from full texts did not attract much attention. In 2014, the TAC evaluation conference in Knowledge Base Population (KBP) track included so-called sentiment track [6]. The task was to find all the cases where a query entity (sentiment holder) holds a positive or negative sentiment about another entity (sentiment target). Thus, this task was formulated as a query-based retrieval of entity-sentiment from relevant documents and focused only on query entities[1].

MPQA 3.0 [5] is a corpus of analytical articles with annotated opinion expressions (towards entities and events). The annotation is sentence-based. For example, in the sentence «When the Imam issued the fatwa against Salman Rushdie for insulting the Prophet...», Imam is negative to Salman Rushdie, but is positive to the Prophet. The current corpus consists of 70 documents. In total, sentiments towards 4,459 targets are labeled.

In paper [4], authors studied the approach to the recovery of the documents attitudes between subjects mentioned in the text. The approach considers such features as frequency of a named entity in the text, relatedness between entities, direct-indirect speech, etc. The best quality of opinion extraction obtained in the work was only about 36% F-measure, which shows that the necessity of improving extraction of attitudes at the document level is significant and this problem has not been sufficiently studied.

For the analysis of sentiments with multiple targets in a coherent text, in the works [2,14] the concept of sentiment relevance is discussed. Ben-Ami et al. [2] consider several types of thematic importance of the entities discussed in the text: the main entity, an entity from a list of similar entities, accidental entity, etc. These types are treated differently in sentiment analysis of coherent texts.

Each attitude may be considered in terms of related article contexts. The context consists of words and may be treated as an *embedding*, where each word represents a vector of features. Convolving such embeddings by a set of different filters, in paper [17] authors implemented and trained the Convolutional Neural

---

[1] https://tac.nist.gov/2014/KBP/Sentiment/index.html.

Network (CNN) model for the relation classification task. Being applied for the SemEval-2010 Task 8 dataset [7], the obtained model significantly outperformed the results of other participants.

This idea was further proceeded in terms of *max pooling* operation [16]. This is an operation, which is applied to the convolved by filters data and extracts the maximal values within each convolution. However, for the relation classification task, original max pooling reduces information extremely rapid, and hence, blurs significant relation aspects. Authors proposed to treat each convolution in parts. The division into parts was related to attitude entities and was as follows: *inner* (between entities), and *outer*. This approach results in an advanced architecture model and was dubbed as «Piecewise Convolutional Neural Network» (PCNN).

3   Corpus and Annotation

We use RuSentRel v1.0 corpus[2] consisted of analytical articles from Internet-portal inosmi.ru [10]. These articles in the domain of international politics are obtained from foreign authoritative sources and translated into Russian. The collected articles contain both the author's opinion on the subject matter of the article and a large number of attitudes mentioned between the participants of the described situations.

For the documents, the manual annotation of the sentiment attitudes towards the mentioned named entities have been carried out. The annotation divided into two subtypes:

1. The author's relation to mentioned named entities;
2. The relation of subjects expressed as named entities to other named entities.

These opinions were recorded as triples: (*Subject of opinion, Object of opinion, attitude*). The attitude can be negative (*neg*) or positive (*pos*), for example, (*Author, USA, neg*), (*USA, Russia, neg*). Neutral opinions or lack of opinions are not recorded. Attitudes are described for the whole documents, not for each sentence. In some texts, there were several opinions of the different sentiment orientation of the same subject in relation to the same object. This, in particular, could be due to the comparison of the sentiment orientation of previous relations and current relations (for example, between Russia and Turkey). Or the author of the article could mention his former attitude to some subject and indicate the change of this attitude at the current time. In such cases, it was assumed that the annotator should specify exactly the current state of the relationship. In total, 73 large analytical texts were labeled with about 2000 relations.

To prepare documents for automatic analysis, the texts were processed by the automatic name entity recognizer, based on CRF method [11]. The program identified named entities that were categorized into four classes: Persons, Organizations, Places and Geopolitical Entities (states and capitals as states). In total, 15.5 thousand named entity mentions were found in the documents of the collection.

---

[2] https://github.com/nicolay-r/RuSentRel/tree/v1.0.

An analytical document can refer to an entity with several variants of naming (*Vladimir Putin – Putin*), synonyms (*Russia – Russian Federation*), or lemma variants generated from different wordforms. Besides, annotators could use only one of possible entity's names describing attitudes. For correct inference of attitudes between named entities in the whole document, corpora provides the list of variant names for the same entity found in our corpus. The current list contains 83 sets of name variants. This allows separating the sentiment analysis task from the task of named entity coreference.

Table 1. Statistics of RuSentRel v1.0 corpus

Parameter	Training collection	Test collection
Number of documents	44	29
Avg. number of sentences per doc.	74.5	137
Avg. number of mentioned NE per doc.	194	300
Avg. number of unique NE per doc.	33.3	59.9
Avg. number of positive pairs of NE per doc.	6.23	14.7
Avg. number of negative pairs of NE per doc.	9.33	15.6
Avg. dist. between NE within a sentence in words	10.2	10.2
Share of attitudes expressed in a single sentence	76.5%	73%
Avg. number of neutral pairs of NE per doc.	120	276

A preliminary version of the RuSentRel v1.0 corpus was granted to the Summer school on Natural Language Processing and Data Analysis[3], organized in Moscow in 2017. The collection was divided into the training and test parts. In the current experiments we use the same division of the data. Table 1 contains statistics of related parts of the RuSentRel corpus. The last line of the Table shows the average number of named entities pairs mentioned in the same sentences without indication of any sentiment to each other per a document. This number is much larger than number of positive or negative sentiments in documents, which additionally stresses the complexity of the task.

4 Contexts

In this paper, the task of sentiment attitude extraction is treated as follows: given an attitude as a pair of its named entities, we predict a sentiment label of a pair, which could be positive, negative, or neutral. The act of extraction is to select only those pairs, which were predicted as non-neutral.

The main assumption of sentiment attitude existence between a pair of entities is their relatively short distance in the text. We consider a *context* as a

---

[3] https://miem.hse.ru/clschool/.

text fragment that is limited by a single sentence and includes a pair of named entities. It allows us to cover up to 76% attitudes for train collection and 73% for test respectively (see Table 1). Consider context c related to the attitude a, when both object and subject (or their synonyms) of a appear in c. Therefore, for each attitude we have a set of related contexts.

Separately for train and test collections, we compose and group these sets by sizes and the resulted statistics for the first eight groups is presented in Table 2. We decide a context *sentiment* with a pair of entities, when related sentiment attitude could be found, and *neutral* otherwise. Irrespective of a context label, in most cases we deal with single-context attitudes in train and test collections. However, the distribution of the sentiment single-context attitudes represent 48%, – is about a half of all occured attitudes. Considering such a distinctive factor for attitudes labeling, it is important to take into account the labels of several contexts.

Table 2. Context-based attitude representation statistics for RuSentRel-v1.0

Parameter	Total 100%	1	2	3	4	5	6	7	8
Train-neutral	5008	81%	6.7%	1.6%	0.5%	0.2%	0.1%	0.2%	0.1%
Train-sent	456	48%	16%	3.4%	4.3%	2.0%	0.9%	0.9%	0.9%
Test-neutral	8002	79%	6.6%	2.0%	1.0%	0.5%	0.3%	0.3%	0.2%
Test-sent	655	47%	13%	5.0%	3.5%	2.5%	1.3%	1.2%	1.3%

5   Context Classification

For context label prediction, we use an approach that does not depend on hand-crafted features. An advanced CNN model proposed by [16] has been implemented. Besides architecture details and transformation aspects, Fig. 1 illustrates a dataflow for an attitude with «USA» and «Russia» as named entities in following context: «... *The US is considering the possibility of new sanctions against Russia* ... ». Next, we describe each architectural block in details.

5.1   Context Embedding

Each context unit, i.e. word, entity, punctuation sign, number and so on, is treated as a *term*. Section 5.6 provides verbosely description towards text processing aspects. Each context has been centered by the middle term between subject and object and limited by k terms.

Let E_t is a precomputed embedding vocabulary for terms. Given i'th term t of context, we compose its embedding vector as a concatenation of:

1. $E_t(t) \in \mathbb{R}^{l_w}$ – term embedding vector, where l_w is a vector size;
2. $E_p(t) \in \mathbb{R}^{l_p}$ – term part-of-speech embedding vector;
3. d_1 and $d_2 \in \mathbb{R}^{l_d}$ – distance to object and subject entities;

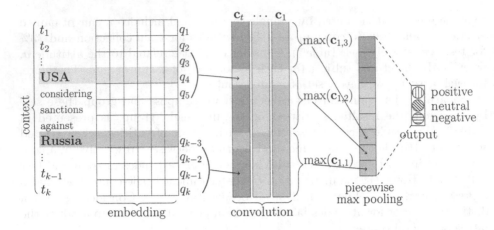

Fig. 1. Piecewise Convolutional Neural Network

Given an attitude entity e_1, let $d_1 = E_p(pos(t) - pos(e_1))$, where $pos(\cdot)$ is a position index in sample s by a given argument, and $E_d \in \mathbb{R}^{k \times l}$ represents a distance embedding matrix. The same computations are applied for d_2 with the other entity e_2 respectively. Finally, composed $E_{ctx} \in \mathbb{R}^{k \times m}$ represents an embedded context.

5.2 Convolution

This step of data transformation applies filters towards the context embedding. Treating the latter as a feature-based attitude representation, this approach implements feature merging by sliding a filter of a fixed size within a dataset and transforming information in it.

We regard E_{ctx} from Sect. 5.1 as a sequence of rows $Q = \{\mathbf{q}_1, \ldots, \mathbf{q}_k\}$, where $\mathbf{q}_i \in \mathbb{R}^m$. We denote $\mathbf{q}_{i:j}$ as consequent vectors concatenation from i'th till j'th positions. An application of $\mathbf{w} \in \mathbb{R}^d$, $(d = w \cdot m)$ towards the concatenation $\mathbf{q}_{i:j}$ is a sequence *convolution* by filter $\mathbf{w}$, where w is a filter window size. Figure 1 illustrates $w = 3$. For convolving calculation c_j, we apply scalar multiplication as follows:

$$c_j = \mathbf{w}\mathbf{q}_{j-w+1:j} \tag{1}$$

Where $j \in \overline{1 \ldots k}$ is filter offset within the sequence Q. We decide to let $\mathbf{q}_i$ a zero-based vector of size m in case when $i < 0$ or $i > k$. As a result, $\mathbf{c} = \{c_1, \ldots, c_k\}$ with shape $\mathbf{c} \in \mathbb{R}^k$ is a convolution of a sequence Q by filter w.

To get multiple feature combinations, a set of different filters $W = \{\mathbf{w}_1, \ldots \mathbf{w}_t\}$ has been applied towards the sequence Q, where t is an amount of filters. This leads to a modified Formula 1 by introduced layer index i as follows:

$$c_{i,j} = \mathbf{w}_i\mathbf{q}_{j-w+1:j} \tag{2}$$

Denoting $c_i = \{c_{i,1}, \ldots, c_{i,n}\}$ in Formula 2, we reduce the latter by index j and compose a matrix $C = \{c_1, c_2, \ldots, c_t\}$ which represents convolution matrix with shape $C \in \mathbb{R}^{k \times t}$. Figure 1 illustrates an example of convolution matrix with $t = 3$.

5.3 Piecewise Max Pooling

Max pooling is an operation that reduces values by keeping maximum. In original CNN architecture, max pooling applies separately per each convolution $\{c_1, \ldots, c_t\}$ of t layers. It reduces convolved information quite rapidly, and therefore is not appropriate for attitude classification task. To keep context aspects that are inside and outside of the attitude entities, we perform *piecewise max pooling*. Given attitude entities as borders, we divide each c_i into inner, left and right segments $\{c_{i,1}, c_{i,2}, c_{i,3}\}$. Then max pooling applies per each segment separately:

$$p_{i,j} = max(c_{i,j}), \quad i \in \overline{1 \ldots t}, \quad j \in \{1, 2, 3\} \tag{3}$$

Thus, for each c_i we have a set $p_i = \{p_{i,1}, p_{i,2}, p_{i,3}\}$. Concatenation of these sets $p_{i:j}$ results in $p \in \mathbb{R}^{3t}$ and that is a result of piecewise max pooling operation. At the last step we apply the hyperbolic tangent activation function. The shape of resulted d remains unchanged:

$$d = tanh(p), \in \mathbb{R}^{3t} \tag{4}$$

5.4 Sentiment Prediction

To receive a context sentiment predictions, the result $d \in \mathbb{R}^{3t}$ of max pooling operation passed through a pair of fully connected hidden layers $\langle W_1, b \rangle$:

$$o = W_1 d + b_1, \qquad W_1 \in \mathbb{R}^{c \times 3t}, \quad b_1 \in \mathbb{R}^c \tag{5}$$

Where h denotes a first layer output size, c is an expected amount of classes, and o is an output vector. The elements of the latter vectors are unscaled values. We use a *softmax* transformation to obtain probabilities per each output class. Figure 1 illustrates $c = 3$. To prevent a model from overfitting, we employ *dropout* for output neurons of Formula 5 during the training process.

5.5 Training

As a function, the implemented neural network model state θ depends on input parameters set, and hidden parameters which are trainable during network optimization.

Given a list of m contexts, the input represents a set of samples $\{s_1, \ldots, s_m\}$, where $s_i = \langle E_{ctx}, y \rangle$ includes context embedding E_{ctx} and related label $y \in \mathbb{R}^c$. Hidden state consist of the following parameters: E_p, E_d, W, W_1, b.

The neural network training process organized as follows

1. Composing a set of bags $I = \{b_1, \ldots, b_N\}$ where $b_i = \{s_1, \ldots, s_g\}$;
2. Performing a forward propagation through the network and receive set of output samples $O = \{o_1, \ldots, o_q\} \in \mathbb{R}^{q \times c}$, where $q = g \cdot N$;
3. Computing *cross entropy loss* for output as follows:

$$l_i = \sum_{j=1}^{c} \log p(y_i | o_{i,j}; \theta), \; i \in \overline{1 \ldots q} \qquad (6)$$

4. Composing *cost* as maximal loss within each group of bags; from previous step:

$$cost_i = max\{l_{i*g} \; \ldots \; l_{(i+1)*g}\}, \; i \in \overline{1 \ldots N} \qquad (7)$$

5. Using *cost* to update hidden variables set;
6. Repeat steps 2–5 while the necessary epochs count will not be reached.

5.6 Text Processing and Embedding Details

All context information were converted intro terms. Table 3 provides an example of text transformation. Starting with *words*, i.e. text parts that separated by spaces, which could be converted into other text units:

1. **Tokens**, in case of the beginning or ending of a word related to a list of pre-defined tokens[4], i.e punctuation signs, numbers; related parts were departed from word;
2. **Entities**, which could be a single or concatenation of words.

As for tokens embedding, we use randomly implemented embedding. For words and entities we use E_w embedding vocabulary. In the latter case $\mathbf{e}_{w_i} = E_w(w_i)$[5]. Additionally, each w_i has been lowercased and lemmatized before accessing the E_w. Attitude entities considered as a single words or multiple in case of phrases (for example «Russian Federation»). The embedding values for them is an average embedding of related words.

Table 3. Example of processing context into list of terms

Context	"President **Putin** notes it in interview": – vice-president **Dmitry Rogozin** wrote this in an article in 2012
Terms	[«"», President, «**Putin**», notes, it, in, interview, «"», «:», «–», vice-president, «**Dmitry Rogozin**», wrote, this, in, an, article, in, «NUMBER», «.»]

---

[4] https://github.com/nicolay-r/sentiment-erc-core/tree/release_19_1.

[5] We use the zero vector value in case of a word absence in E_w.

6 Extracting Sentiment Attitudes

For each attitude a, given a list of extracted and classified contexts, the task is to compose an aggregated sentiment label. In this paper we examine an application of the following functions:

1. L_{first} – keeping the label of only the first appeared context [13];
2. L_{avg} – keeping an average label by voting;
3. $L_{rnn}, L_{gru}, L_{lstm}$ – an application of recurrent neural network models.

Recurrent neural networks (RNN) allows us to consequentially analyze contexts. These class of neural networks consists of hidden state h, with an optional output for a variable length input sequence $X = \{x_1, \dots, x_n\}$, passed step by step. Let x_t is an element of sequence X at step t, the hidden state updates as follows:

$$h_t = f(h(t-1), x_t) \tag{8}$$

In formula 8, f is an nonlinear activation function in neural network, which could be as simple as elementwise logistic sigmoid function. The model based on such architecture is denoted in this paper as L_{rnn}. The activation function f also could be a complex as long-short term memory (LTSM) unit [8], or gated recurrent unit (GRU) proposed by [3] in 2014. We use L_{lstm} and L_{gru} for the models based on LSTM and GRU architectures respectively.

We utilize outputs of the sentiment context classification model, described in Sect. 5, to predict a sentiment of the aggregated attitude. The context position within the article was also taken into account. Providing input as a sequence of embedded contexts[6], each context consists of:

1. $o \in \mathbb{R}^c$ – context predictions (see Sect. 5.4);
2. $p \in \mathbb{R}^{l_s}$ – is a one-hot vector of context position in text, with size of l_s;

To compose p, we consider a text in l_s parts. Let i is an index of a text fragment which the context is belongs to. We set $p[i] = 1$ and thereby emphasize the context position within the text.

7 Experiments

As a measure of classification quality, we take the averaged Precision, Recall and F-measure of positive and negative classes. All the extracted contexts (see Table 2) should be classified as having positive, negative, or neutral sentiment from one named entity of the pair (opinion holder) to the second entity of the pair (opinion target). To extract sentiment attitude from related contexts, we apply various functions discussed in Sect. 6.

Table 4 illustrates the predefined settings of a PCNN model[7] for sentiment context extraction. Starting with a model input, each minibatch has N bags,

---

[6] We treat the contexts in order of their appearance in the text.
[7] https://github.com/nicolay-r/sentiment-pcnn/tree/ccis-2019.

where each bag consisted of g samples. All contexts were limited by k terms. The convolutional filter size w and the *filters count* were chosen according to the prior experiments [13]. We use the *adadelta* optimizer for model training with parameters that were chosen according to [15]. For dropout probability, the statistically optimal value ρ for most classification tasks was chosen.

Table 4. Predefined training parameters

Minibatch $\langle N, g \rangle$	k	w	Filters count	Adadelta params $\langle lr, \rho, \epsilon \rangle$	Dropout ρ
$\langle 6, 3 \rangle$	50	3	300	$\langle 0.1, 0.95, 10^{-6} \rangle$	0.5

Table 5. Embedding parameters

l_w	l_p	l_d	Words in E_t	Entities in E_t	Tokens in E_t	Window size of E_w	Words found in E_w
1000	5	5	147 358	1 417	17	20	63%

Settings of all used embeddings described in Table 5. Parameters l_t, l_p, l_d were used to denote term, part-of-speech, and distance vector embedding sizes respectively. We use a precomputed embedding model for *words* E_w to compose a model for terms E_t by expanding E_w with *tokens* and *entities* (see Sect. 5.6). We utilize Yandex Mystem[8] both for text lemmatization and part-of-speech tagging. The choice of E_w depended on the average distance between entities within a context. According to Table 1, we were interested in a Skip-gram with window size parameter greater than 10 terms. We use a precomputed and publicly available Word2Vec model[9] based on news articles with *window size* of 20.

We have reproduced test with the same label calculation method as in [13], which this paper is related to L_{first}. Applying the model during the training process towards the train collection, it reaches maximum F-measure 85% with accuracy parameter value of 98%, and further optimization leads to results variation on test collection. For each label aggregation method, the F-measure variation on the test collection illustrated in Fig. 2. Mean values are illustrated in Table 6.

We conclude a general significance of usage L_{avg} calculation method comparing with L_{first}, according to significant difference in mean $F_1(P, N)$ (see Fig. 2). For recurrent neural networks, we apply model with settings presented in Table 7. Parameter a_c denotes the limit of contexts per attitude. We use smaller value of a_c than possible maximum to prevent models from forgetting context positions history.

---

[8] https://tech.yandex.ru/mystem/.

[9] http://rusvectores.org/static/models/rusvectores2/news_mystem_skipgram_ 1000_20_2015.bin.gz.

Table 6. Experiment results

Methods	Precision	Recall	F-measure
PCNN$_{first}$ [13]	0.37	0.26	0.31
PCNN with L_{rnn}	0.37	0.29	0.32
PCNN with L_{lstm}	0.37	0.30	0.32
PCNN with L_{gru}	0.37	0.29	0.32
PCNN with L_{avg}	0.37	0.32	0.33
KNN	0.18	0.06	0.09
Naïve Bayes (Gauss)	0.06	0.15	0.11
Naïve Bayes (Bernoulli)	0.13	0.21	0.16
SVM (Default values)	0.35	0.15	0.15
SVM (Grid search)	0.09	0.36	0.15
Random forest (Default values)	0.44	0.19	0.27
Random forest (Grid search)	0.41	0.21	0.27
Gradient boosting (Default values)	0.36	0.06	0.11
Gradient boosting (Grid search)	0.47	0.21	0.28
Experts agreement	0.62	0.49	0.55

Table 7. Recurrent Neural Network parameters

Batch size	a_c	Epochs	Unit types	Hidden size
2	10	2	{rnn, lstm, gru}	128

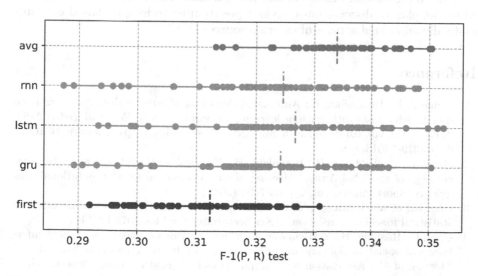

Fig. 2. Result test collection F-measure variation of each label aggregation function during training process; dashed lines shows mean value.

According to Fig. 2, switching *unit type* from rnn leads to non significant improvements, especially because of a limited amount of contexts were used for training. Overall, comparing recurrent neural network-based label aggregation methods with L_{first} and L_{avg}, we conclude an importance to consider labels of all related contexts, not only first appeared (Table 6).

The proposed approach significantly outperforms feature-dependent conventional approaches [10]. Comparing with the proposed approach, in [10] each attitude is presented as a single vector of handcrafted features. Using the same dataset, SVM and Naive Bayes achieved 16% F-measure, and the best result has been obtained by the Random Forest classifier (27% F-measure). Overall, we conclude that this task still remains complicated and the results are quite low. It should be noted that in [4] authors worked with much smaller documents written in English and reported F-measure 36%.

8   Conclusion

This paper introduces the problem of sentiment attitude extraction from mass-media articles written in Russian. The model based on Convolutional Neural Networks were used to extract sentiment attitudes' contexts. A variety of aggregation functions based on contexts were described. Such functions were used to define a attitude sentiment by its contexts.

In the current experiments, the problem of sentiment attitude extraction is considered as a three-class machine learning task. The proposed CNN based model in combination with label aggregation functions perform better than conventional classifiers.

Due to the dataset limitation and manual annotating complexity, in further works we plan to discover unsupervised pre-training techniques based on automatically annotated articles of external sources.

References

1. Alimova, I., Tutubalina, E.: Automated detection of adverse drug reactions from social media posts with machine learning. In: van der Aalst, W., et al. (eds.) AIST 2017. LNCS, vol. 10716, pp. 3–15. Springer, Cham (2018). https://doi.org/10.1007/978-3-319-73013-4_1
2. Ben-Ami, Z., Feldman, R., Rosenfeld, B.: Entities' sentiment relevance. In: Proceedings of the 52nd Annual Meeting of the Association for Computational Linguistics, Short Papers, vol. 2, pp. 87–92 (2014)
3. Cho, K., et al.: Learning phrase representations using RNN encoder-decoder for statistical machine translation. arXiv preprint arXiv:1406.1078 (2014)
4. Choi, E., Rashkin, H., Zettlemoyer, L., Choi, Y.: Document-level sentiment inference with social, faction, and discourse context. In: Proceedings of the 54th Annual Meeting of the Association for Computational Linguistics, Long Papers, vol. 1, pp. 333–343 (2016)

5. Deng, L., Wiebe, J.: MPQA 3.0: an entity/event-level sentiment corpus. In: Proceedings of the 2015 Conference of the North American Chapter of the Association for Computational Linguistics: Human Language Technologies, pp. 1323–1328 (2015)
6. Ellis, J., Getman, J., Strassel, S., M.: Overview of linguistic resources for the TAC KBP 2014 evaluations: planning, execution, and results. In: Proceedings of TAC KBP 2014 Workshop, National Institute of Standards and Technology, pp. 17–18 (2014)
7. Hendrickx, I., et al.: Semeval-2010 task 8: multi-way classification of semantic relations between pairs of nominals. In: Proceedings of the Workshop on Semantic Evaluations: Recent Achievements and Future Directions, pp. 94–99. Association for Computational Linguistics (2009)
8. Hochreiter, S., Schmidhuber, J.: Long short-term memory. Neural Comput. 9(8), 1735–1780 (1997)
9. Loukachevitch, N., Rubtsova Y., V.: Sentirueval-2016: overcoming time gap and data sparsity in tweet sentiment analysis. In: Computational Linguistics and Intellectual Technologies Proceedings of the Annual International Conference Dialogue, Moscow, RGGU, pp. 416–427 (2016)
10. Loukachevitch, N., Rusnachenko, N.: Extracting sentiment attitudes from analytical texts. In: Proceedings of International Conference of Computational Linguistics and Intellectual Technologies Dialog-2018 (2018)
11. Mozharova, V.A., Loukachevitch, N.V.: Combining knowledge and CRF-based approach to named entity recognition in Russian. In: Ignatov, D., et al. (eds.) AIST 2016. CCIS, vol. 661, pp. 185–195. Springer, Cham (2017). https://doi.org/10.1007/978-3-319-52920-2_18
12. Rosenthal, S., Farra, N., Nakov, P.: Semeval-2017 task 4: sentiment analysis in Twitter. In: Proceedings of the 11th International Workshop on Semantic Evaluation (SemEval-2017), pp. 502–518 (2017)
13. Rusnachenko, N., Loukachevitch, N.: Extracting sentiment attitudes from analytical texts via piecewise convolutional neural network (2018). ceur-ws.org
14. Scheible, C., Schütze, H.: Sentiment relevance. In: Proceedings of the 51st Annual Meeting of the Association for Computational Linguistics, Long Papers, vol. 1, pp. 954–963 (2013)
15. Zeiler, M.D.: ADADELTA: an adaptive learning rate method. arXiv preprint arXiv:1212.5701 (2012)
16. Zeng, D., Liu, K., Chen, Y., Zhao, J.: Distant supervision for relation extraction via piecewise convolutional neural networks. In: Proceedings of the 2015 Conference on Empirical Methods in Natural Language Processing, pp. 1753–1762 (2015)
17. Zeng, D., Liu, K., Lai, S., Zhou, G., Zhao, J.: Relation classification via convolutional deep neural network. In: Proceedings of COLING 2014, the 25th International Conference on Computational Linguistics: Technical Papers, pp. 2335–2344 (2014)

Discovering, Classification, and Localization of Emergency Events via Analyzing of Social Network Text Streams

Dmitriy Deviatkin[2(✉)], Artem Shelmanov[1,2], and Daniil Larionov[2,3]

[1] Skolkovo Institute of Science and Technology, Moscow, Russia
[2] Federal Research Center "Computer Science and Control"
of Russian Academy of Sciences, Moscow, Russia
devyatkin@isa.ru
[3] People's Friendship University of Russia, Moscow, Russia

Abstract. We present text processing framework for discovering, classification, and localization emergency related events via analysis of information sources such as social networks. The framework performs focused crawling of messages from social networks, text parsing, information extraction, detection of messages related to emergencies, automatic novel event discovering, matching them across different sources, as well as event localization and visualization on a geographical map. For detection of emergency-related messages, we use CNN and word embeddings. The components of the framework are experimentally evaluated on Twitter and Facebook data.

Keywords: Event detection · Topic modelling · Monitoring ·
Named entity recognition · Text processing · Novel topic

1 Introduction

Recent research showed that Twitter, Facebook, and other social networks have valuable applications in emergency situations. Since large-scale emergency events give rise to a massive publication activity in social networks [41], these resources accumulate information about situation in affected areas, infrastructure damage, casualties, requests and proposals for help. They have already been used for enhancing situation awareness of affected people and emergency response teams [3,17,24], as well as for online detecting and monitoring emergency events like earthquakes [32,34]. Advanced information retrieval techniques can detect emergencies in text streams automatically so direct appeals to the rescue services through the standard channels may not be needed.

This research continues the previous studies presented in [10–12] that are devoted to monitoring restricted geographical regions via social networks for enhancing situation awareness during emergency situations. In this work, we

© Springer Nature Switzerland AG 2019
Y. Manolopoulos and S. Stupnikov (Eds.): DAMDID/RCDL 2018, CCIS 1003, pp. 180–196, 2019.
https://doi.org/10.1007/978-3-030-23584-0_11

solve the task of automatic discovering and classification of emergency events via parsing stream of text messages. We consider an event in a text stream as a group of topically related messages that reflect a real-life event in a small time period. Since we are looking for emergency events, it is crucial to detect them as soon as possible: long before they become trendy and gain high amount of publications. Therefore, one of the peculiarities of this task is the problem of identification of novel topics that correspond to emergency events. It is also important to distinguish events (earthquakes, fire breakouts, storms, hurricanes, etc.) that happen in different locations at the same time despite they generate topically similar text streams (e.g., destructions caused by a single storm that moves across a country should be identified as different events).

We present a multimodal topic model for event discovering that leverages spatial information, as well as describe approaches to assessing event novelty and matching events from different information sources. We evaluate several models for detection of emergency related messages based on various types of embeddings and classification techniques including deep learning. We present a multimodal topic model for event discovering that leverages spatial information, as well as describe approaches to assessing event novelty and matching events from different information sources. We also propose methods for event classification and localization on a geographical map. The experimental evaluations on collections of messages from Twitter and Facebook show that our methods outperform the baselines.

This paper extends [11] with a review of state-of-the-art methods for event classification (Sect. 2), with description and experimental evaluation of the developed methods for ecology-related emergency event classification (Sects. 3.4 and 5.2), and with a technique for event localization and visualization on a geographical map. The rest of the paper is structured as follows. Section 2 reviews the related work. Section 3 describes the natural language pipeline of our system including the subsystem for extraction and classification of emergency related messages. Section 4 presents the developed method for novel emergency event discovering and matching across information sources. Section 5 describes the experimental evaluation of methods. Section 6 concludes and outlines the future work.

2   Related Work

The work related to our current research includes publications considering the tasks of event detection in microblogs, topic evolution tracking, as well as emerging topic detection. Most of the approaches to these problems can be divided into two major groups.

The first group of methods for emerging event detection and tracking primarily relies on topic models adopted to temporal aspects of the task. They are based on different modifications of PLSA models [14]. One of the fundamental works in this area is [7]. It proposes several dynamic topic models that align topics across time steps with logistic normal distribution, train with approximation based on variational Kalman filters and perform inference with the help of wavelet regression. Another fundamental model named "topics over time" is presented in [36]. Authors propose a method for jointly modelling both word co-occurrences and localization in continuous time without employing Markov assumption. Another topic model that takes into account temporal dimension is on-line LDA presented in [1]. In this approach, distributions generated on the previous time steps are used as priors for word generation on the current step. For each topic, the method builds transformation matrix that captures the evolution of the topic over time. Authors consider a topic as emerging if it is significantly differs from topics in the same time period or from all topics seen before. For topic comparison, Kullback-Leibler divergence is used. In [37], researchers instead of creating monolith Bayesian model propose to learn a topic model and a transition matrix to shift distributions over discrete time steps. They formulate the problem of model learning as minimizing the least square error between predicted topic distribution using transformation and the actual topic distribution of new documents. The proposed approach provides the ability to predict topic trends in the future. Other notable related work on topic models for emerging topic detection in microblog data include Twitter-LDA [13], BBTM (bursty biterm topic model) [40], and TopicSketch [39].

The second group of methods is based on detection of emerging features like terms, keywords, or token segments, and clustering of them. In [8], to define emerging terms authors use two metrics named "nutrition" and "energy function" (biology metaphor). Nutrition of a term is calculated as a sum of modified term frequency in a tweet multiplied by author importance (calculated via PageRank) summed through all tweets in a time period. The energy function of a term is proportional to the difference of its current nutrition and its nutrition in the previous time intervals. Authors declare a term as emerging if its energy value is more than a "critical drop" value, which is proportional to the average energy of all terms in the current time period. Using co-occurrence of terms, authors build a graph with edges that correspond to the strongest relationships between terms. The emerging terms become seeds of strongly connected components that finally represent emerging topics. Authors of [38] consider frequencies of words as signals and decode these signals with wavelet analysis to find emerging words. Some trivial words are filtered out by analyzing their corresponding signal autocorrelations. The remaining words are then clustered to form events with a modularity-based graph partitioning technique. In [33], the emerging keywords are identified using significance measure based on outlier detection algorithm. More specifically, authors used exponentially weighted average of terms and co-occurring terms. For detection of novel events, in [22], researchers propose to use instead of single unigrams so called "event segment" – key phrases for an event that possibly refer to named entities or semantically meaningful information units. They cluster event segments into events considering both their frequency distribution and content similarity. Emerging segments are detected by abnormal frequency distribution of the tweet and user frequencies of the segments. Importance of an event is also determined by Wikipedia. Authors consider segments that frequently appear as anchors in Wikipedia more favorable.

This approach is intended for finding the most realistic events and to derive the most newsworthy segments to describe the identified events.

The method presented in [15] combines two aforementioned approaches: it uses topic modelling in conjunction with models for emerging terms detection. Topic models are used to detect topic distributions in each time interval. Term novelty is estimated by local weighted linear regression. In order to advance from detection of term novelty to detection of topic novelty, authors solve optimization problem. The solution gives novelty and fading probabilities for a topic. Based on these two probabilities, topic evolution operations are defined subsequently to identify emerging topics from the large number of latent ones and track how these topics evolve over time. To compare topics, authors use Jensen-Shannon distance.

Another approach to emergency event detection employ dictionary learning method [19]. The dictionary contains topics, which consist of atoms (numerical vectors). Vector representation of documents can be approximated with a linear combination of such atoms. The method has two steps: determining novel documents in a text stream and identifying a cluster structure among the novel documents. In the first step, the method checks whether a new document can be represented as a sparse linear combination of known atoms with low error. If it is not the case, the document is considered novel. Such documents are used to learn a new dictionary of novel topics. On the second step, the learned dictionary is used to build clusters of similar novel messages. These clusters are considered as emerging topics.

Our approach to novel event discovering is based on multimodal topic modeling and takes into account spatial information. Its key benefits compared to the previous work are the following.

- It allows to separate similar emergency events happened in different locations (for example, storms or typhoons).
- It provides an obvious way to match messages from different sources (social networks) taking into account location information.
- It can help to reveal location information of an event from a set of scattered messages.

There are few studies devoted to event classification. In [44], an unsupervised topic model is used to classify incident-related messages and events appeared in Twitter. Experimental evaluation of the model is performed on a collection of messages related to recent events in New York City, the Chelsea explosion, and Hurricane Sandy. The results show that the model can extract emergency events and classify them for both small and large-scale events, and hyperparameters of the model can be shared in a similar language environment. The main disadvantage of the proposed method lies in the fact that the model requires large corpora of documents to provide satisfactory performance. Authors note that the event types with lower than 5% covered tweets are difficult to identify. To overcome this problem, the researchers suggest to use lists of keywords obtained by training supervised classification models.

A supervised event classification approach is proposed in [4]. In this work researchers present a framework for creation of emergency detection systems based on the "human as a sensor" (HaaS) paradigm. They define an ontology for the HaaS paradigm in the context of emergency detection. The researchers consider the following types of messages: a trusted message, a primary message, an emergency message. They suggest extracting emergency-type specific knowledge from a large and structured set of messages and using this corpus for statistical analysis and hypothesis testing, checking occurrences or validating filtering within a specific emergency type as a part of the framework. Authors also apply Weka framework to train a decision tree model, which is feasible to use as the fine-grained classifier filter at this stage of the system since most of the noisy messages are filtered out by previous components. They use emergency-specific training set for earthquakes that contains more than 1.4K messages. It is separated into two subsets: messages related and messages not related to a seismic event in progress. During the offline phase, messages of the training set were manually classified by annotators. In [5], researchers propose a classifier, which can differentiate between the relevant tweets of a predefined type and all other types of tweets. It also can detect several types of tweets, which describe different aspects of an event but do not refer to the real event happening at that moment, such as tweets referring to the past events or the messages from the news agencies referring to the events that happened in other regions. They focused on the detection of the tweets that have their geolocation and that were posted by individuals witnessing actual events of a particular type. To identify such relevant tweets, they constructed a predictive model that determines whether a tweet posted at a particular moment was relevant. This model is based on the random sample of messages from existing dataset that were labeled with some the standard machine-learning classification algorithms, such as logistic regression, naive Bayes, or random forest. They use a large set of hand-crafted features extracted from message texts. Some of them are the standard features, such as the length of the message (in characters), the number of words and sentences in the tweet, the presence of a URL link, and so on. They also develop an additional set of features that are focused on the emergency event detection problem and help us to get a better classification performance. Some examples of those features include location of the key concept words in a message, presence of a name of some geographic object, phone number or particular pronouns, percentage of capital letters in the message. Results of a comprehensive study of different features for event type detection are provided in [28]. In this study, researchers investigate the relative importance of different feature types as well as the effect of several feature selection methods. Since the task of detecting mass emergencies is characterized by high heterogeneity of the data, they focus on detecting the features that would be capable of dividing emergency reports from other messages, irrespective of the type of the disaster.

It is worth to note, that in the most of studies related to this topic researchers use complex feature engineering and simple supervised models like SVM or Logistic Regression. We believe that this is due to the small size of labeled datasets for event type detection. However, recent regularization techniques (e.g., dropout) for more complex models, such as convolutional neural networks have made them applicable to this task too.

3 Natural Language Processing Pipeline

Our method for event discovering needs complex preprocessing of natural language texts. We perform basic linguistic analysis, named entity recognition, time recognition, and detection of emergency related texts.

3.1 Basic Linguistic Analysis

The basic linguistic analysis includes tokenization, sentence splitting, postagging, lemmatization, and syntax parsing. The pipeline is implemented via IsaNLP[1] - a library that organizes various NLP components for English and Russian. Tokenization, sentence splitting, postagging, and lemmatization are performed by components based on NLTK toolkit [6]. The syntax parsing is performed by SyntaxNet McParseface [2].

3.2 Named Entity Recognition

We perform extraction of the following types of objects: person's names, organizations, geographical locations, and ship names. For basic NER extraction, we use Polyglot framework. This system uses distant supervision on Wikipedia for learning underlying model and is able to perform named entity recognition for 40 languages. However, we note that performance of such an approach is not suitable for location extraction due to lack of recall. To improve the recall of location extraction and achieve the ability to normalize extracted textual information into geographic coordinates, in the previous work, we implemented a rule- and dictionary-based module [10]. We created a gazetteer from Geonames[2] and supplied it with several filtering rules based on postags of extracted tokens. Geonames also provides mapping of locations into the geographic coordinates.

To extract and normalize temporal expressions, we use a combination of two tools: spaCy[3] (NLP framework based on deep learning) and a datetimeparser[4] (a library based on a set of hand-crafted rules).

For extraction of ship names, in the previous work [12], we implemented a hybrid approach. On the basis of a database of ship names, we implemented a gazetteer that has high recall but low precision due to the fact that many generic words appear to be ship names. To mitigate this problem, we also trained a neural network based on C-LSTM architecture [43]. The network filters out erroneous cases generated by the gazetteer and drastically improves precision and overall F1-score of ship name detection.

---

[1] https://github.com/IINemo/isanlp.
[2] http://www.geonames.org/.
[3] https://spacy.io/.
[4] https://github.com/scrapinghub/dateparser.

3.3 Detection of Emergency Related Messages

For detection of emergency related tweets, we also use a combination of a gazetteer and a neural network. The gazetteer is based on the CrisisLex lexicon, proposed in [26]. This gazetteer generates many false positives that are filtered out by the neural network. For training, we use a CrisisLex corpora [27] and a annotated corpus of messages collected from Twitter. For detection of emergency-related messages, we explore:

- Various embeddings: word-level: fastText [18] (trained on our own corpus/pretrained on English Wikipedia), GloVe [29] (Common Crawl with dimension 300/Twitter with dimension 200), Word2Vec [25], sentence-level: InferSent [9].
- Various types of models: logistic regression (from scikit-learn), random forest (from scikit-learn), gradient boosting on decision trees (LigthGBM algorithm [20]), fully-connected network (FCN), convolutional neural network (CNN), and C-LSTM.

For logistic regression, random forest, gradient boosting algorithms, as well as for FCN we average word embeddings and use the result vector as features. Word-level embeddings in C-LSTM and CNN are processed in a standard way. Sentence-level embeddings are not used in C-LSTM and CNN since these architectures work only with sequences. For the rest algorithms, sentence-level embeddings are used as common features.

The fully-connected network is a simple 2-layer perceptron regularized by dropout. The first layer activation function is ReLU, the outputs of the last layer are passed through the softmax. The architecture of the convolutional neural network for sentence classification was proposed in [21]. In this architecture, padded sequence of word embeddings is processed by a one-dimensional convolution layer, followed by max pooling layer to reduce dimensionality. The result vectors are stacked into a single one and are fed into a fully-connected layer that makes a prediction. Activation functions for convolutional and fully-connected layers are set to ReLU and softmax respectively. The architecture of C-LSTM consists of 1-d convolution layer with ReLU activation and max pooling followed by a LSTM recurrent layer. The final predictions are made by two dense layers with hyperbolic tangent and softmax activations.

3.4 Event Classification

To determine the type of emergency event, we employed additional message classification model. Messages are classified as related to one of the events from the predefined set. In this paper, we focus on ecology-related events, thus the embeddings/models were examined for performance on a set of ecology-related messages from CrisisLexT26 [27]. In addition to the classification methods used for emergency event detection, here, we have also explored the CatBoost algorithm [30].

4   Emergency Event Detection Method

In the first step, we collect all messages from Twitter using topic search API [12] and crisis-related lexicon. Then, we detect emergency related messages among crawled tweets using methods described in Sect. 3.3 and filter out all irrelevant tweets.

In the second step, we train multimodal topic model to identify emergency events described by messages and then determine novel events among them by comparing term distributions of the events from adjacent time periods.

In the third step, we use event-related and location-related lexis from the obtained topics to crawl messages from other sources (Facebook in particular). Then, we apply emergency detection method again and filter out all irrelevant posts. The trained topic model is used to check whether the remaining messages are topically similar to the events extracted from Twitter.

4.1   Identification of Events

In the first step, we discretize the timeline into small time periods (one day in the experiments). In each time period, multimodal topic model with additive regularization [16] is trained.

Let D be a collection of tweets from a time period, let Def be a default modality (regular event-related lexis) and let Loc be a modality devoted to location of events. The main reason to use such modalities is to separate similar events happened in different places in one period of time. We consider each message $d \in D$ as a set of tokens, related to those modalities $W = W \cup W_{loc}$. The goal of the topic modeling is to find factorization of a matrix of empirical probabilities for documents and tokens:

$$\hat{p}(w \vee d) \approx p(w \vee d) = \sum_{t \in T} p(w|t)p(t|d) = \sum_{t \in T} \phi_{wt}\theta_{td}, \forall w \in W \qquad (1)$$

This problem could be solved by maximizing the weighted sum of the following log-likelihoods with additive regularizers:

$$L(\Phi, \Theta) = \sum_{\gamma \in \Gamma} \gamma \sum_{d \in D} \sum_{w \in W_\gamma} n_{dw} ln \sum_{t \in T} \phi_{wt}\theta_{td} + \alpha R_{sp}(\Theta) + \beta R_{sm}(\Phi)$$
$$+ \tau R_{dcorr}(\Phi_{loc}) \to \max_{\Phi, \Theta}. \qquad (2)$$

Here $\gamma \in \Gamma = \{\gamma, \gamma_{loc}\}$ are weights of the modalities, Φ is a matrix of token probabilities for topics, and Θ is a matrix of topic probabilities for documents. As in [16], we apply smooth-sparse regularizers to achieve smooth term distributions in topics and sparse topic distributions in messages:

$$R_{sm}(\Phi) = \sum_{t \in T} KL(\beta_t \vee \phi_{wt}), \qquad (3)$$

$$R_{sp}(\Theta) = -\sum_{d \in D} KL(\alpha_d \vee \theta_{td}), \tag{4}$$

where α_d and α_t are sampled from some predefined distributions.

We apply decorrelation regularizer only for location modality to be able to detect similar events happened in different places at the same time:

$$R_{dcorr}(\Theta_l oc) = -\sum_{t,s \in T} \sum_{w \in W_l oc} \phi_{wt} \phi_{ws}. \tag{5}$$

We use BigARTM library [35] to train multimodal topic models. The result is Φ and Θ matrices for each time period. After that, "background" topics with high entropy of token distributions can be filtered.

4.2   Detection of Novel Events

In the second step, we determine whether the extracted events were discussed before. We aggregate several adjacent periods of time into "time window". Consider we have topics s and t in the same time window. Denote vectors of token distributions for these topics as Φ_t and Φ_s. As in [15], we use Jensen-Shannon divergence between token probabilities for the topics to estimate topic similarity:

$$JSD(\Phi_t, \Phi_s) = \frac{1}{2}(KL(\Phi_t||M)) + KL(\Phi_s||M)) \tag{6}$$

$$M = \frac{1}{2}(\Phi_t + \Phi_s). \tag{7}$$

A topic is denoted as a "new event" if there is no earlier similar topics in a predefined time window.

4.3   Events Matching

In the third step, we match messages related to the same event from different sources, which can be various types of social networks or mass media sites. In the experiments, we enriched messages from Twitter related to novel emergency events with Facebook public posts. For each novel event, we construct a search query as a combination of default and location tokens with the highest weights. To crawl Facebook, we use Ghost.py[5] library.

We filter obtained posts (leaving only emergency related messages) as described in Sect. 3.3 and extract named entities and locations from them. We infer topic-probabilities matrix $\widetilde{\Theta}$ for remaining posts using the pretrained model for the event. Then, we filter all messages, which are not topically similar to the event. Due to the use of multimodal models, information about locations is also taken into account when assessing the similarity of posts.

---

[5] https://github.com/jeanphix/Ghost.py.

5   Experiments

5.1   Detection of Emergency Related Messages

Dataset and Pre-processing. For evaluation of method for detection of emergency related messages, we use the CrisisLexT6 dataset. The dataset consists of 60,000 tweets related to 6 major crisis situations. Emergency related tweets are labeled as "on-topic" and others are labeled as "off-topic". The pre-processing procedure included elimination of the special characters, as well as conversion of hashtags, emojis, and URLs into single tokens.

Hyperparameters. Logistic regression. Regularization: L2 penalty. Tolerance: 1e−4. Inverse regularization strength: 1.0.

Random Forest. Number of estimators: 1,000. No limits to maximum number of features and tree depth. Split quality measure: Gini impurity. Min number of samples per split: 2. Min number of samples per leaf: 1.

Gradient boosting. Maximum tree depth: 20. Number of leaves: 11. Learning rate: 0.05. Feature fraction 0.9. Bagging fraction: 0.8. Min frequency: 5. Number of estimators: 4,000 with early stopping for 200.

FCN. Size of hidden layer: 256. Dropout: 0.5. Number of epochs: 10. Loss: cross entropy. Optimization algorithm: Adam. Learning rate: 1e−4, no decay. Batch size: 256.

CNN. Kernel size: [3, 4, 5]. Number of filters: 512. Dropout: 0.5. Optimization algorithm: Adam. Learning rate: 1e−4. Loss: binary cross entropy. Batch size: 128. Vocabulary size: 10,001. Number of epochs: 10 with early stopping for 3 epochs.

Results and Discussion. We use 5-fold cross-validation for evaluation. Results are presented in Table 1. We discovered several insights into problems with processing and analyzing crisis and Twitter specific lexicon:

Table 1. Results of the models for emergency-related message detection (F1-score), %

Models	FstTrain	FstWiki	GloVeCC	GloVeTwt	Word2Vec	InferSent
Log Reg	87.4 ± 8.4	82.5 ± 9.2	88.6 ± 5.3	85.1 ± 6.9	88.9 ± 6.7	89.4 ± 4.9
Rnd For	86.9 ± 9.5	82.3 ± 11.1	87.4 ± 7.4	83.9 ± 10.5	87.4 ± 8.9	89.4 ± 4.9
GBDT	91.7 ± 0.1	89.8 ± 0.1	93.0 ± 0.1	89.8 ± 0.2	92.0 ± 0.2	N/A
FCN	90.9 ± 0.3	89.8 ± 0.1	92.2 ± 0.3	88.0 ± 0.2	91.2 ± 0.3	90.8 ± 0.2
CNN	**94.3 ± 0.3**	93.4 ± 0.3	93.8 ± 0.2	92.7 ± 0.2	92.9 ± 0.2	N/A
CLSTM	92.1 ± 0.2	92.2 ± 0.3	92.2 ± 0.6	91.5 ± 0.5	92.3 ± 0.5	N/A

- Sentence-level embeddings are better than averaging word vectors. Averaging embeddings of all words in a tweet blur the real meaning of text. InferSent embedding model, which is constructed using NLI data and BiL-STM encoders, treats sentence as a single entity and performs more general projection process. But the higher dimensionality (required to make accurate projections) makes it harder to use several classification algorithms.
- GloVe embeddings pretrained on a Common Crawl corpus show better results than Twitter specific embeddings. Sentence-level embeddings, pretrained on non-specific natural language inference data, also show superior results. It seems reasonable that crisis-related lexicon differs from common Twitter lexicon and tends to be closer to common lexicon. However, we should note that there is a lack of available Twitter data for training. GloVe Twitter corpus contains only 27 billion words, which is much less compared to Common Crawl corpus of 840 billion words.
- All neural network models have lower standard deviation of F1-score compared to other machine learning algorithms (except GBDT). Therefore, the quality of neural networks could be much stable on unseen data and less sensitive to the context.
- Our best classifier (CNN for text classification + fastText, trained on our dataset) outperforms models presented in the related work [23,31,42].

5.2   Event Classification

Dataset and Pre-processing. For training and evaluation of event classification method we use the data from CrisisLexT26 [27]. The dataset contains tweets from 26 large emergency events in 2012–2013, labeled by informativeness. The ecology-related subset of tweets was created by randomly selecting 3,000 messages, related to ecology events (Singapore Haze, Australia and Colorado wildfires, Venezuela refinery explosion) and 3,000 messages related to other events. Selected messages were marked as positive and negative examples respectively. Pre-processing was the same as before: special characters elimination and hashtags/emojis conversion.

Hyperparameters. Most of the hyperparameters are the same as in previous section, with two exceptions. In Yoon Kim CNN, we reduced number of convolutional kernels (256) and increased number of epochs (50) with the same early stopping regularization; in CLSTM we increased number of epochs (50). For CatBoost, we empirically choose the following hyperparameters: loss function: logloss, number of iterations: 500, l2 coef.: 3, subsample rate: 0.66, tree depth: 6.

Results and Discussion. Results are presented in Table 2. They are slightly different from the previous task. However, several ideas are still valid for that results:

- Ecology-related crisis lexicon also differs from common Twitter lexicon. That is why better results are shown by models that use GloVe embeddings, pre-trained on Common Crawl corpora.
- Deep Neural Networks outperforms other types of classification algorithms. Indeed, while emergency detection task is solved better by convolutional network, CLSTM shows 95% F1-score for ecology-related messages.

Table 2. Results of the models for event classification (F1-score), %

Models	FstTrain	FstWiki	GloVeCC	GloVeTwt	Word2Vec
Log Reg	64.58 ± 0.62	80.55 ± 1.72	90.11 ± 1.72	79.55 ± 2.09	62.73 ± 0.88
Rnd For	80.38 ± 0.68	80.89 ± 2.04	88.70 ± 1.37	81.14 ± 1.17	73.02 ± 1.04
GBDT-CatB	80.72 ± 1.43	81.98 ± 0.14	89.66 ± 0.73	81.44 ± 0.94	72.71 ± 1.27
GBDT-LGBM	83.10 ± 0.88	82.85 ± 0.60	90.74 ± 0.66	82.96 ± 0.50	76.22 ± 1.53
FCN	66.50 ± 1.39	81.40 ± 1.47	90.82 ± 1.73	81.79 ± 1.79	62.93 ± 1.78
CNN	75.43 ± 1.72	94.55 ± 0.38	93.67 ± 1.24	93.85 ± 0.97	74.40 ± 2.04
CLSTM	77.77 ± 1.19	92.71 ± 0.86	**95.16 ± 0.66**	91.96 ± 1.33	76.22 ± 0.91

5.3 Novel Emergency Event Extraction

Dataset and Pre-processing. We crawled 60k Twitter messages from April 1, 2018 to April 12, 2018 using the focused crawler presented in [12]. With the help of CNN neural network, we filtered out messages that are not related to emergency events, which reduced the number of tweets in the dataset to 5,200. The remaining tweets were analyzed with the natural language processing pipeline and with the event discovery method. After that, we also crawled Facebook posts for each extracted event. Using the developed method, we filtered out posts that were considered irrelevant to events extracted from Twitter. After filtering, 1,000 Facebook posts left.

Hyperparameters. In our experiments, we applied grid search to tune weights of the regularizers for topic models. A criterion for the search was a weighted sum of model perplexity, model's matrices sparsity and model's pointwise mutual information.

Results and Discussion. Since the experiments were conducted on open data, we estimated only precision of models. The results are presented in Table 3. The experiment shows that the proposed approach outperforms baseline LDA models. This confirms the importance of using information about the locations in the framework. One can note relatively low precision for the events matching. We believe this is due to substantial lag of time between the message crawling and the event matching experiments. Thus, true event-related posts may be treated by Facebook's search as less actual than others.

Table 3. Results of the novel emergency event extraction method (Precision), %

Step	LDA (baseline)	Multimodal model
All events	63.3	93.3
Novel events	71.4	80.0
Event matching	60.0	67.0

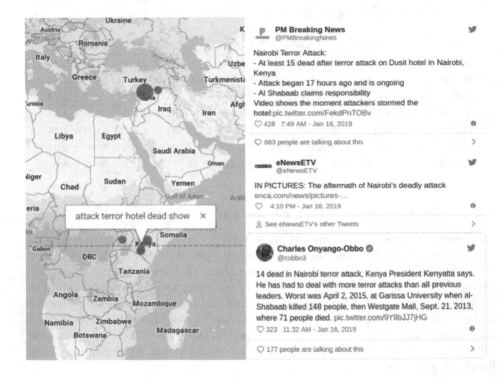

Fig. 1. Event visualization on a geographical map

5.4   Event Localization and Visualization

To present the results of our research, we created the web service[6], which controls Twitter crawling, topic modeling process, and visualizes emergency events on map (Fig. 1. Event location is determined by top-3 extracted tokens from "Location" modality of the topic model. These tokens are transformed to latitude-longitude by online geocoding service. Then, coordinates are used to place marker of a specific size on map. Size is calculated using number of messages that belong to each event. Also, we show some tweets related to event under clicked marker.

---

[6] http://nlp.isa.ru/ru/demo/emergency-messages.

6 Conclusion

We considered several problems related to monitoring of social networks: detection of messages related to emergencies, extraction of novel events, event classification, and matching events reflected in different text sources. For detection and classification of emergency-related messages, we use CNN and word embeddings. For extraction and localization of novel events and matching them across different sources, we propose a multimodal topic modelling enriched with spatial information and Jensen-Shannon divergence.

We investigated the performance of different algorithms and embeddings for emergency-related message detection on CrisisLexT6 dataset and found that the best solution is given by CNN with fastText embeddings. We also compared the proposed multimodal topic model and the LDA baseline. The experimental results are promising and show that the proposed framework could be useful for monitoring emergency events via messages in social media.

Acknowledgements. The project is supported by the Russian Foundation for Basic Research, project numbers: 16-29-12901, 15-29-06045 "ofi_m".

References

1. AlSumait, L., Barbará, D., Domeniconi, C.: On-line LDA: adaptive topic models for mining text streams with applications to topic detection and tracking. In: 2008 Eighth IEEE International Conference on Data Mining, ICDM 2008, pp. 3–12. IEEE (2008)
2. Andor, D., et al.: Globally normalized transition-based neural networks. In: Proceedings of the 54th Annual Meeting of the Association for Computational Linguistics, Long Papers, vol. 1, pp. 2442–2452 (2016)
3. Ashktorab, Z., Brown, C., Nandi, M., Culotta, A.: Tweedr: mining Twitter to inform disaster response. In: Proceedings of ISCRAM, pp. 354–358 (2014)
4. Avvenuti, M., Cimino, M.G., Cresci, S., Marchetti, A., Tesconi, M.: A framework for detecting unfolding emergencies using humans as sensors. SpringerPlus 5(1), 43 (2016)
5. Bauman, K., Tuzhilin, A., Zaczynski, R.: Using social sensors for detecting emergency events: a case of power outages in the electrical utility industry. ACM Trans. Manag. Inf. Syst. (TMIS) 8(2–3), 7 (2017)
6. Bird, S., Klein, E., Loper, E.: Natural Language Processing with Python: Analyzing Text with the Natural Language Toolkit. O'Reilly Media, Inc., Sebastopol (2009)
7. Blei, D.M., Lafferty, J.D.: Dynamic topic models. In: Proceedings of the 23rd International Conference on Machine Learning, pp. 113–120. ACM (2006)
8. Cataldi, M., Di Caro, L., Schifanella, C.: Emerging topic detection on twitter based on temporal and social terms evaluation. In: Proceedings of the Tenth International Workshop on Multimedia Data Mining, p. 4. ACM (2010)
9. Conneau, A., Kiela, D., Schwenk, H., Barrault, L., Bordes, A.: Supervised learning of universal sentence representations from natural language inference data. In: Proceedings of the 2017 Conference on Empirical Methods in Natural Language Processing, pp. 670–680 (2017)

10. Deviatkin, D., Shelmanov, A.: Towards text processing system for emergency event detection in the arctic zone. In: Proceedings of Data Analytics and Management in Data Intensive Domains, pp. 225–232 (2016)
11. Deviatkin, D., Shelmanov, A., Larionov, D.: Discovering novel emergency events in text streams. In: Proceedings of Data Analytics and Management in Data Intensive Domains, pp. 208–215 (2018)
12. Devyatkin, D., Shelmanov, A.: Text processing framework for emergency event detection in the arctic zone. In: Kalinichenko, L., Kuznetsov, S.O., Manolopoulos, Y. (eds.) DAMDID/RCDL 2016. CCIS, vol. 706, pp. 74–88. Springer, Cham (2017). https://doi.org/10.1007/978-3-319-57135-5_6
13. Diao, Q., Jiang, J., Zhu, F., Lim, E.P.: Finding bursty topics from microblogs. In: Proceedings of the 50th Annual Meeting of the Association for Computational Linguistics: Long Papers, vol. 1, pp. 536–544. Association for Computational Linguistics (2012)
14. Hofmann, T.: Probabilistic latent semantic indexing. In: Proceedings of the 22nd Annual International ACM SIGIR Conference on Research and Development in Information Retrieval, pp. 50–57. ACM (1999)
15. Huang, J., Peng, M., Wang, H., Cao, J., Gao, W., Zhang, X.: A probabilistic method for emerging topic tracking in microblog stream. World Wide Web **20**(2), 325–350 (2017)
16. Ianina, A., Golitsyn, L., Vorontsov, K.: Multi-objective topic modeling for exploratory search in tech news. In: Filchenkov, A., Pivovarova, L., Žižka, J. (eds.) AINL 2017. CCIS, vol. 789, pp. 181–193. Springer, Cham (2018). https://doi.org/10.1007/978-3-319-71746-3_16
17. Imran, M., Castillo, C., Lucas, J., Meier, P., Vieweg, S.: AIDR: artificial intelligence for disaster response. In: Proceedings of the Companion Publication of the 23rd International Conference on World Wide Web Companion, pp. 159–162 (2014)
18. Joulin, A., Grave, E., Bojanowski, P., Mikolov, T.: Bag of tricks for efficient text classification. In: Proceedings of the 15th Conference of the European Chapter of the Association for Computational Linguistics, vol. 2, pp. 427–431 (2017)
19. Kasiviswanathan, S.P., Melville, P., Banerjee, A., Sindhwani, V.: Emerging topic detection using dictionary learning. In: Proceedings of the 20th ACM International Conference on Information and Knowledge Management, pp. 745–754. ACM (2011)
20. Ke, G., et al.: LightGBM: a highly efficient gradient boosting decision tree. In: Advances in Neural Information Processing Systems, pp. 3149–3157 (2017)
21. Kim, Y.: Convolutional neural networks for sentence classification. In: Proceedings of the 2014 Conference on Empirical Methods in Natural Language Processing (EMNLP), pp. 1746–1751 (2014)
22. Li, C., Sun, A., Datta, A.: Twevent: segment-based event detection from tweets. In: Proceedings of the 21st ACM International Conference on Information and Knowledge Management, pp. 155–164. ACM (2012)
23. Li, H., Caragea, D., Caragea, C., Herndon, N.: Disaster response aided by tweet classification with a domain adaptation approach. J. Contingencies Cris. Manag. **26**(1), 16–27 (2018)
24. MacEachren, A.M., et al.: SensePlace2: GeoTwitter analytics support for situational awareness. In: Proceedings of the IEEE Conference on Visual Analytics Science and Technology (VAST), pp. 181–190 (2011)
25. Mikolov, T., Sutskever, I., Chen, K., Corrado, G.S., Dean, J.: Distributed representations of words and phrases and their compositionality. In: Advances in Neural Information Processing Systems, pp. 3111–3119 (2013)

26. Olteanu, A., Castillo, C., Diaz, F., Vieweg, S.: CrisisLex: a lexicon for collecting and filtering microblogged communications in crises. In: Proceedings of ICWSM (2014)
27. Olteanu, A., Vieweg, S., Castillo, C.: What to expect when the unexpected happens: social media communications across crises. In: Proceedings of the 18th ACM Conference on Computer Supported Cooperative Work and Social Computing, pp. 994–1009. ACM (2015)
28. Pekar, V., Binner, J., Najafi, H., Hale, C.: Selecting classification features for detection of mass emergency events on social media. In: Proceedings of the International Conference on Security and Management (SAM), p. 192. The Steering Committee of The World Congress in Computer Science, Computer Engineering and Applied Computing (2016)
29. Pennington, J., Socher, R., Manning, C.: Glove: global vectors for word representation. In: Proceedings of the 2014 Conference on Empirical Methods in Natural Language Processing (EMNLP), pp. 1532–1543 (2014)
30. Prokhorenkova, L., Gusev, G., Vorobev, A., Dorogush, A.V., Gulin, A.: CatBoost: unbiased boosting with categorical features. In: Advances in Neural Information Processing Systems, pp. 6639–6649 (2018)
31. Roy Chowdhury, S., Purohit, H., Imran, M.: D-sieve: a novel data processing engine for efficient handling of crises-related social messages. In: Proceedings of the 24th International Conference on World Wide Web, pp. 1227–1232. ACM (2015)
32. Sakaki, T., Okazaki, M., Matsuo, Y.: Earthquake shakes Twitter users: real-time event detection by social sensors. In: Proceedings of the 19th International Conference on World Wide Web, pp. 851–860. ACM (2010)
33. Schubert, E., Weiler, M., Kriegel, H.P.: SigniTrend: scalable detection of emerging topics in textual streams by hashed significance thresholds. In: Proceedings of the 20th ACM SIGKDD International Conference on Knowledge Discovery and Data Mining, pp. 871–880. ACM (2014)
34. Unankard, S., Li, X., Sharaf, M.A.: Emerging event detection in social networks with location sensitivity. World Wide Web 18(5), 1393–1417 (2015)
35. Vorontsov, K., Frei, O., Apishev, M., Romov, P., Dudarenko, M.: BigARTM: open source library for regularized multimodal topic modeling of large collections. In: Khachay, M.Y., Konstantinova, N., Panchenko, A., Ignatov, D.I., Labunets, V.G. (eds.) AIST 2015. CCIS, vol. 542, pp. 370–381. Springer, Cham (2015). https://doi.org/10.1007/978-3-319-26123-2_36
36. Wang, X., McCallum, A.: Topics over time: a non-Markov continuous-time model of topical trends. In: Proceedings of the 12th ACM SIGKDD International Conference on Knowledge Discovery and Data Mining, pp. 424–433. ACM (2006)
37. Wang, Y., Agichtein, E., Benzi, M.: TM-LDA: efficient online modeling of latent topic transitions in social media. In: Proceedings of the 18th ACM SIGKDD International Conference on Knowledge Discovery and Data Mining, pp. 123–131. ACM (2012)
38. Weng, J., Lee, B.S.: Event detection in Twitter. In: ICWSM 2011, pp. 401–408 (2011)
39. Xie, W., Zhu, F., Jiang, J., Lim, E.P., Wang, K.: TopicSketch: Real-time bursty topic detection from Twitter. IEEE Trans. Knowl. Data Eng. 28(8), 2216–2229 (2016)
40. Yan, X., Guo, J., Lan, Y., Xu, J., Cheng, X.: A probabilistic model for bursty topic discovery in microblogs. In: AAAI, pp. 353–359 (2015)

41. Yin, J., Karimi, S., Robinson, B., Cameron, M.: ESA: emergency situation aware-
    ness via microbloggers. In: Proceedings of the 21st ACM International Conference
    on Information and Knowledge Management, pp. 2701–2703. ACM (2012)
42. Zhang, S., Vucetic, S.: Semi-supervised discovery of informative tweets during the
    emerging disasters. arXiv preprint arXiv:1610.03750 (2016)
43. Zhou, C., Sun, C., Liu, Z., Lau, F.: A C-LSTM neural network for text classifica-
    tion. arXiv preprint arXiv:1511.08630 (2015)
44. Zuo, F., Kurkcu, A., Ozbay, K., Gao, J.: Crowdsourcing incident information for
    emergency response using open data sources in smart cities. Transp. Res. Rec.
    2672(1), 198–208 (2018)

Citation Content Analysis and a Digital Library

Sergey Parinov[1,2](✉) [iD]

[1] Central Economics and Mathematics Institute of RAS, Moscow, Russia
sparinov@gmail.com
[2] Russian Presidential Academy of National Economy and Public
Administration (RANEPA), Moscow, Russia

Abstract. This paper presents an approach of two-way data exchange between the citation content analysis, provided by the Cirtec project, and the big research digital library Socionet. Many papers in Socionet have citation relationships with other papers and also linkages with authors' personal profiles and through them with other information objects. It allows making an enrichment of data for the citation content analysis by different additional information and, as well, linking results of such analysis with objects in a digital library, like papers, their authors, affiliation organizations, etc. We discuss what numeric and qualitative indicators can be built by citation content analysis based on the Cirtec open citation data. Since these indicators have IDs related with digital library objects, they can be integrated and visualized as computer-generated annotations to papers' full texts in PDF.

Keywords: Citation content analysis · Cirtec/CyrCitEc · Digital library · Socionet · Integration · Indicators

1 Introduction

Currently there is a clear trend in the research community to make more re-usable citation data from research papers. One of illustrations is the OpenCitations project. The main aim of this project is "the creation and current expansion of the Open Citations Corpus (OCC), an open repository of scholarly citation data made available under a Creative Commons public domain dedication, which provides in RDF accurate citation information (bibliographic references) harvested from the scholarly literature" (http://opencitations.net/). As of January 20, 2019, the OCC has ingested the references from 326,743 citing bibliographic resources and contains information about 13,964,148 citation links to 7,565,367 cited resources.

The focus of the OpenCitation project is the references. Another part of citation data that also available in research papers is the citation content or context. Waltman [19] in his review of the traditional citation impact indicators proposed different ways for the indicators' improvement, including taking into account "the context in which a publication is referenced (i.e., the sentences in a citing publication around the reference to a cited publication)" [19], p. 43.

© Springer Nature Switzerland AG 2019
Y. Manolopoulos and S. Stupnikov (Eds.): DAMDID/RCDL 2018, CCIS 1003, pp. 197–211, 2019.
https://doi.org/10.1007/978-3-030-23584-0_12

In recent years, methods for analyzing the content of citations have been actively developed. Some studies [8] present a concept of the content-based citation analysis (CCA), which addresses a citation's value. "The text of citation context is used to characterize publications for various applications, such as publication summarization, survey article generation and information retrieval" [9]. Other authors wrote: "the extraction of citation contexts is a preliminary step to any statistical, distributional, syntactic or semantic analysis" [5]. Also "To capture document usage, we observe that the context in which one document cites another tends to reflect how a document is used, namely, within a document, people tend to cite other documents for very precise reasons" [1].

Practical experiments with the extraction and analysis of the in-text citations (they are also called as the in-text references) on various sets of full text papers are also known. One of them identified verbs in citation contexts [2]. "Our hypothesis is that the semantic meaning of the relation that exists between the cited work and the citing article is often expressed, to some extent, by the verb phrase in the sentence containing the in-text reference" [5]. These authors also characterized the different sections of articles in terms of the verbs that appear in citation contexts [3].

Another aspect of the citation content analysis is how references are distributed along the structure of articles, as well as the age of these cited references [4]. Some authors analyzed in-text citations as functions of time, textual progression, and scientific field. They built characteristics of the in-text citations in over five million full text articles [7].

Hernández-Alvarez and Gómez [10] in their survey of citation context analysis provided information about used tasks, techniques, and resources, including such tasks as the citation polarity and function classifications.

The analysis of citation polarity/function has a potential for conclusions with some accuracy about the motives of authors to cite papers. Such analysis can also produce suggestions: what exactly from the cited papers and for what purposes were used in the citing papers. In some cases, this information may be critically important to authors of the cited papers and may help to initiate direct communication between them and citing authors.

The second section of the paper presents the Cirtec (former CyrCitEc) project and its current results, which aimed to provide an open citation content data source.

In the third section, we describe some services of the digital library Socionet that use the citation content data. One of them gives better representation of citations in papers' full text PDF. Another, a network of semantic linkages among information objects in Socionet and its API create an opportunity to enrich citation content data by additional information.

The last section discusses the question: what indicators could be built by the citation content analysis taking into account that the indicators must be useful to enrich traditional representation of papers, authors and organizations in a DL? We present a set numerical and qualitative indicators for which there are the first experimental calculations.

This paper is an extended version of the paper "Open Citation Content Data and a Digital Library" published at CEUR in the proceedings of the XX International Conference "Data Analytics and Management in Data Intensive Domains" [15].

Comparing this paper with its shorter version, we added some corrections and updates; and also discussion of possible set of indicators which can be built using the unique citation data provided by the Cirtec project.

2 Open Citation Content Data

One of few already existed sources of open citation content data is the In-text Reference Corpus (InTeReC) available at https://zenodo.org/record/1203737. Currently the InTeReC dataset provides 314,023 sentences containing in-text references (also called as the in-text citations) together with other useful data. The sentences are extracted from 90,071 research articles published by PLOS5 up to September 2013 [5].

A full text of each sentence in InTeReC is supplemented by [5]:

- a journal title;
- DOI of the article from which the sentence was extracted;
- size of the article, as number of sentences, and a position of the sentence in the article, as number of sentences from the beginning of the article;
- size of the section, as number of sentences, and a position of the sentence in the section, as number of sentences from the beginning of the section;
- section type (introduction, method, results, etc.);
- a list of verb phrases that occur in the sentence.

Another source of open citation content data is provided by our ongoing project Cirtec (https://github.com/citeccyr). The project is funded by the Russian Presidential Academy of National Economy and Public Administration (RANEPA, http://www.ranepa.ru/eng/).

The project has two main aims: (1) to create a public service for processing available research papers full text (particularly, in PDF and with main focus on Social Sciences), in order to build and regularly update an open dataset of citation relationships and citations content; (2) to use the citation content data for developing methods of qualitative citation analysis, which can be used for improving a current practice of a research performance assessment.

The project tends to provide a pilot version of open scholarly infrastructure [6] based on following pillars:

- Open distributed architecture. It means providing a concept, open source soft-ware and an initial core infrastructure for interoperable systems, which are processing citation relationships and content from research papers' full text. Two initial nodes of this core infrastructure, presented by interacting CitEc (http://citec.repec.org/) and Cirtec systems. Currently these nodes are exchanging by citations data. The nodes have a specialization on processing papers in specific languages: Romano-Germanic languages by CitEc and Russian by Cirtec. Other nodes, e.g. specialized on processing citation data in languages, like Chinese, Japanese, Arabic, etc., can be added by the same way. There is also an intention to integrate data about references into the OpenCitations Corpus (http://opencitations.net/).
- Transparency. It allows publishers, authors and readers of papers to see for each paper how their citation data are created by the system and to trace why some

papers' references/in-text citations are not processed or not counted. Better representation and usability of citation data by its deeper integration with a digital library (DL) tools and services.

- Enrichment facilities. The system should provide tools for authors of papers to enter additional data to correct errors while processing citations from their papers and to enrich their citation relationships, e.g. by qualitative characteristics of their motivation for citing papers of other authors, etc.
- Public control. Readers of papers should see how authors used enrichment facilities to increase their number of citations. Public will be able to react on wrong authors behaviour.

Currently Cirtec takes papers' metadata from the Socionet digital library (https://socionet.ru/), which also includes a full set of metadata from RePEc (http://repec.org).

Comparing with InTeReC the Cirtec system has following main differences:

- an openness for adding new papers for processing by the system, the papers just have to be added to a digital library Socionet or into RePEc;
- the system in everyday mode automatically processes all new papers at its in-put and daily updates citation content data;
- the input papers are in PDF (InTeReC works with papers in XML).

In Cirtec we use the term "in-text citation" instead of the "in-text reference" accepted in InTeReC. "The in-text citations of publications are the citations referred to this publication in the full text of other publications cited this publication. The text around the in-text citation is the citation context text" [9].

In the end of January 2019, Cirtec processed 321 collections of papers with about 158,000 publications in total. The biggest part of this set are 253 Russian academic journals covering different academic disciplines and provided by NEICON[1]. There are also research papers series in Russian and English languages provided by RANEPA[2], the Higher School of Economics[3], and some other Russian Universities.

An approach used by Cirtec for citation data parsing was presented in [14]. Victor Lyapunov and Sergey Petrov built all needed software to parse citation data from PDF documents.

All extracted by Cirtec project citation data and processing log files are publicly available at http://cirtec.ranepa.ru/data/. This storage is organized as nested folders with names based on Socionet IDs of processed papers. E.g. such folder contains[4]:

1. JSON version of PDF papers (file 0.pdf-stream.json), which was used for parsing citation data;
2. file "summary.xml" with the parsed citation data; and

---

[1] https://socionet.ru/collection.xml?h=spz:neicon&l=en.

[2] https://socionet.ru/collection.xml?h=repec:rnp&l=en.

[3] https://socionet.ru/collection.xml?h=repec:hig&l=en.

[4] http://cirtec.ranepa.ru/data/RePEc/hig/fsight/v%253A11%253Ay%253A2017%253Ai%253A4%253Ap%253A84-95/.

3. reports about errors in processing the paper and parsing citation data (files with extensions ".err" and ".log").

Daily updated aggregated statistics about parsing results for each collection are available at http://cirtec.ranepa.ru/stats.html. Thomas Krichel maintains these statistics.

Processing statistics of this set of collections show (on the end of January 2019) that only 75% of total papers have full text PDF available for the citation data parsing and only about 50% of total papers have a list of references in more or less standard form.

Based on the subset of papers with references we parsed in total 1,336,363 references that is in average 18 references per paper. In this set, we have about 5% of duplicated references, because different papers can cite the same publications and can have the same references.

For 178,301 of parsed references we were able to create citation relationships between citing and cited papers, since we found cited papers' metadata within Socionet.

Additionally, we parsed 1,260,803 in-text citations. They mention 1,168,885 of parsed references. It is in 167,478 references less than total number of parsed references, since some references are not mentioned at all.

Non-mentioned references were also counted[5]: approximately 167,000 references (it is about 12% of total) have no mentions in the in-text citations at all. About 38% of papers with references have at least one non-mentioned reference.

One in-text citation includes following data (see also an example of the data below):

1. a text string of how this in-text citation is occurred in a paper content, e.g. a number or an author name in square or round brackets (the tag <Exact> in the example below);
2. a link to a reference, mentioned in this in-text citation (the tag <Reference> below);
3. text coordinates of the in-text citation, i.e. a serial number of the first and the last in-text citation symbols counting from the beginning of the paper's content (tags <Start> and <End>);
4. citation contexts located at the left and at the right according the in-text citation; it includes at least 200 symbols expanded for taking a whole sentence (tags <Prefix> and <Suffix>).

An example of parsed data about one in-text citation:

```
<intextref>
<Prefix>...countries and Soviet republics</Prefix>
<Suffix>; Gokhberg, Kuznetsova, 2011]. ...</Suffix>
<Start>8757</Start>
<End>8781</End>
<Exact>[Gokhberg et al., 2009</Exact>
<Reference>20</Reference>
</intextref>
```

Source: https://goo.gl/1FAkCH

---

[5] In [14] we listed types of in-text citations that were processed.

The in-text citation from the example above has a link with a reference having the number 20 in a paper. Cirtec parsed for this reference following data:

```
<reference
  num="20"
  start="54464"
  end="54654"
  author="Gokhberg Kuznetsova ..."
  title="Towards ..."
  year="2009"
  handle="repec:oup:scippl:v:36:y:2009:i:2:p:121-126">
  <from_pdf>Gokhberg L., Kuznetsova T., Zaichenko S.
(2009) Towards a New Role of Universities in Russia: Pro-
spects and Limitations. Science and Public Policy, vol. 36,
no 2, pp. 121-126.
  </from_pdf>
</reference>
```

Source: https://goo.gl/1FAkCH

The XML data of the example above includes following subtags and attributes:

1. subtag `<from_pdf>` - extracted raw data of a reference (some publishers provided reference data within papers' metadata, see as an example any citation data file of NEICON archive);
2. attribute num - a serial number of the reference in the list;
3. attributes `start` and end - text coordinates of the reference, which are numbers of the first and the last symbols of the reference counted from the beginning of the initial PDF document's text;
4. attribute `url` – contains a proper URL, if there is one in data of the tag `<from_pdf>`;
5. attributes author, title and year are extracted from the row reference data in the tag `<from_pdf>` and used for different purposes, e.g. for searching in-text citations by author names, for linking the reference with metadata of the same paper (creating a citation relationship for this reference), etc.;
6. attribute `handle` – contains ID of the paper at Socionet digital library, if the linking procedure for this reference was successful.

These data about in-text citations and references are supplemented by the ID of paper's metadata in a DL (see `<source handle =` in the example below) and by the URL of the source full text PDF of the paper (see `<futli url =` below). Using the paper's metadata ID one can have all available information about this paper, including its title, abstract, authors, etc.

```
<source handle="repec:hig:fsight:v:11:y:2017:i:4:...">
<futli url="https://foresight-journal.hse.ru/data/...">
```

Source: https://goo.gl/1FAkCH

Comparing with InTeReC the Cirtec data source has following main differences:

- the citation content is organized as a text at the right and at the left according the in-text citation location and it provides several sentences instead of one sentence in InTeReC;
- a broader set of attributes for citation content, like reference data linked with in-text citation, etc.
- in-text citation's coordinates as number of symbols (InTeReC counts sentences);
- current version of Cirtec citation data has no associations with the type of paper's sections that exists in InTeReC.

3 Digital Library and Citation Content Data

The citation data provided by the Cirtec project include ID of source papers and URL of their full texts (see the last example above). Such links allow us, using API of a digital library like Socionet [16], to provide new features for DL users and to enrich citation content analysis, as well. In general, it opens new ways to make the knowledge more open.

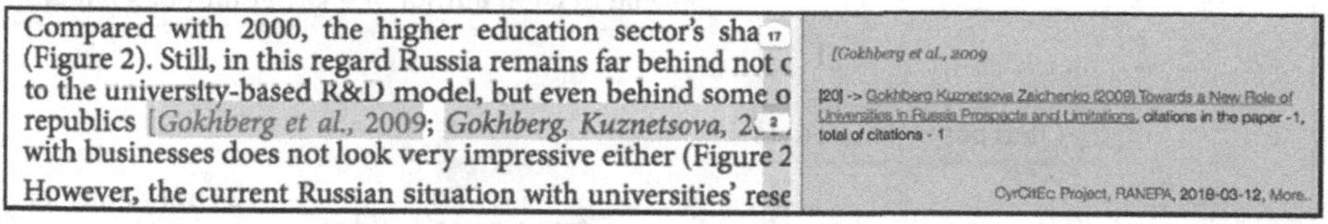

Compared with 2000, the higher education sector's sha ₁₇ (Figure 2). Still, in this regard Russia remains far behind not c to the university-based R&D model, but even behind some o republics [Gokhberg et al., 2009; Gokhberg, Kuznetsova, 2₀₂ with businesses does not look very impressive either (Figure 2 However, the current Russian situation with universities' rese

[Gokhberg et al., 2009

[20] -> Gokhberg Kuznetsova Zaichenko (2009) Towards a New Role of Universities in Russia Prospects and Limitations, citations in the paper -1, total of citations - 1

CyrCitEc Project, RANEPA, 2018-03-12, More.

Fig. 1. An in-text citation as an interactive element (Source: https://goo.gl/bZJwzZ)

Socionet services, as it is described in [14], use in-text citations and references data to produce computer-generated annotations to the content of PDF papers. Figure 1 shows how these annotations look like using the in-text citation and the reference from the examples above.

Readers of PDF papers see the in-text citations, if they exist, as an annotated text. At Fig. 1 there are mentions of two references in brackets. These highlighted in-text citations works as interactive elements, since a click on them opens an information box (at the right side on Fig. 1) with additional data about the cited paper. The additional data can include details about the cited paper, including citing statistics, title, authors, etc.; indicators built on the citation contexts (see its discussion in the next section); links to some tools, etc.

Another Socionet feature is the multiple semantic relationships between information objects [13]. It allows an enrichment of citation content analysis by expanding the citation data with additional attributes belonged to semantically linked information objects. A fragment of the semantic linkage network existed at Socionet is presented at Fig. 2.

Using these linkages, we can associate with the citation data different additional data of cited and citing papers, like titles, authors, classification codes and other elements of their metadata.

Currently, Socionet already has about 70000 authors' profiles (most of them are provided by the RePEc Author service[6]). These authors' profiles have linked with authors' papers, and with about 15000 profiles of organizations (most of them are provided by the RePEc EDIRC service[7]). The organizations' profiles also have links with other authors' profiles belonged to their staff.

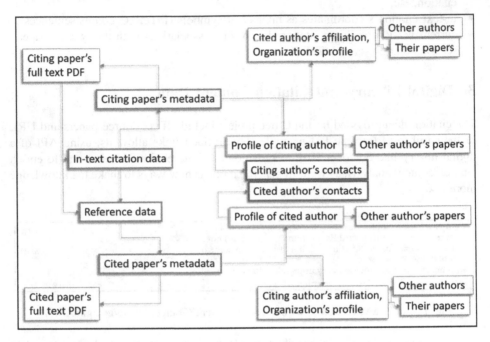

Fig. 2. Semantic linkage network in a digital library, a fragment

Socionet API allows taking different data related with selected information object specified by its ID:

1. Paper's metadata including list of semantically linked objects. The in-text citation and reference examples from the previous section have, as a source, a paper with the ID: repec:hig:fsight:v:11:y:2017:i:4:p:84-95. To take the paper's metadata one should run API: https://socionet.ru/fs/ap.cgi?h=_paper's-ID_[8]. The output is a metadata in XML, which additionally to usual bibliographic data about a paper data includes authors' profile ID (see within the API output the tag <coauthor handle = "repec:per:pers:pku327"...), paper's references (<out><citation>), paper's in-text citations in a form of annotations to full text PDF (<annotations><annot>), and many other data.

---

[6] https://authors.repec.org/.

[7] https://edirc.repec.org/.

[8] An example - https://socionet.ru/fs/ap.cgi?h=repec:hig:fsight:v:11:y:2017:i:4:p:84-95.

2. Author's personal profile data including linked objects, like author's papers, affiliation organization, etc. To take author's profile data, which ID is specified in the example above, run the API: https://socionet.ru/fs/ap.cgi?h=_author's-ID_[9]. It output includes a list of papers' ID, which the author claimed as own, and the author's workplace ID (<workplace><link handle = "repec:edi:aneeeru"...).
3. Organization's profile data including links with authors who work for it. API works here similarly as in previous case.

By this way, one can aggregate various sets of enriched citation data for different scenarios of the citation content analysis. One can collect the citation contents from all papers for a specified author to analyze how the author cite other papers. Similar collection of data can be created to analyze how the author is cited by other researchers. Results of such analysis can be aggregated for groups of authors, e.g. who works for the same organization, etc.

4   Discussion

Using the method from the previous section to enrich initial citation content by metadata of citing and cited papers/authors, we get new opportunities for the citation content analysis. At the same time, we also get an opportunity to return back results of such analysis into DL and link them with DL objects like papers, authors, organizations, journals, etc.

The question is what indicators could be built by the citation content analysis taking into account that the indicators must be useful to enrich traditional representation of papers, authors and organizations in a DL?

The main data source for building the indicators is the citation data provided Cirtec project, including citation contexts from the pairs of tags <Prefix> and <Suffix> for each in-text citation and so on (see the first example in the section "Open citation content data"). As mentioned above we have 1,260,803 records of such citation data.

4.1   Grouping of Citation Contexts

One of options to prepare this dataset for the analysis is to assort in groups the citation contexts having something in commons. The commons can exist, e.g., if the contexts are somehow related with the same DL information objects, like articles citing the same paper, papers of the same author, etc. If we have groups of citation contexts, we can analyze their inter-group characteristics and assign them to the linked DL objects.

The most obvious case is a grouping of the citation contexts linked with the same references (cited papers). Such groups potentially contain many useful information about the cited paper. For instance, they can help to understand: what is similar in a style of how the cited papers are mentioned in citation contexts of citing papers, what relationships exist between cited papers and citing papers, etc.

---

[9] An example - https://socionet.ru/fs/ap.cgi?h=repec:per:pers:pku327.

To make such grouping, one should first recognize the same references among available citation data and then collect as groups the citation contexts citing the unique references. Grouping of the same references is an easy operation when data of each reference includes some common digital code, like DOI. In general, it is not trivial task, since references parsed from the "References" sections of papers belonged to different journals or sources with different styles of their bibliographic representation, or they can just have typos, etc.

Making experiments[10] with grouping of the same references in the Cirtec project we started with comparing the reference text strings by the Damerau–Levenshtein distance[11] algorithm. It works very efficient for recognition references even with few misspellings, but it takes too much time to process the data. In our case, it worked more than 24 h for processing about 1.3 million references.

To make such processing faster we ran a comparison of normalized text strings combined the "author+year+title" data of references. It works much faster, but the result strongly depends on correctness of recognition of these values in reference data and, of course, it fails when there are typos in reference data.

Using our citation dataset we grouped the same references and then also grouped the citation contexts linked with them. We called such groups as "bundles" of citation contexts for the unique references. A bundle got a name "author (year) title" of the initial reference and some unique ID, which allows us to link indicators built for this bundle (see below) with traditional for DL representation of papers, authors and organizations (see the previous section).

Experimentally produced bundles are listed in the left column of the table "Index of References" at http://cirtec.ranepa.ru/analysis/. They are ordered by numbers of papers ("pubs" column), which have this reference in its "References" sections. The biggest "pubs" values are at the top. The list currently presents only 923 bundles having "pubs" value not less than 10.

For each bundle in this list we created the citation data file in XML format available by hyperlink "raw data". The "raw data" file provides information extracted from metadata and full text of papers citing a reference, including the citation contexts and other useful information.

4.2 Quantitative Characteristics

Based on the "raw data" available for bundles we can build at least following numeric indicators[12]:

Frequency of Reference's Mentions. Frequency of reference's mentions in a full text of all papers which have this reference in their "References" sections.

It equals to the number of in-text citations for the reference. It is the most obvious indicator and it has been discussed in many papers, e.g. see a review in [17].

---

[10] Victor Lyapunov made needed software and calculations for these experiments.

[11] https://en.wikipedia.org/wiki/Damerau%E2%80%93Levenshtein_distance.

[12] Thomas Krichel and Roman Puzyrev made needed software and calculations.

This type of indicators means an amount of attention of citing authors to the cited research output. So it can be interpreted as a measure of scholarly impact of the cited paper and its author. It has high significance for the research community, since it can help with improving the traditional citation index. The traditional index counts a number of papers having the selected reference in the papers' "References" sections, but, as it was highlighted above, some references are not mentioned at all in papers' full text. In that case, if papers do not specify in their full text how the reference was used, it can't be recognized as important one for measuring an impact of the cited research output.

We counted the in-text citations within each bundle and presented these numbers in the column "cits" of the table at http://cirtec.ranepa.ru/analysis/. Putting together in this table the traditional citation index (numbers in the column "pubs") and the numbers of mentions for cited papers (the column "cits"), we got a clear evidence of absence of correlation between these two indicators. E.g. five top cited papers by the number of mentions (in the column "cits") have in total 866 mentions and only 137 in total by the column "pubs". The same proportion for the five top papers by traditional citation index is: they have 369 in total for "pubs" and 451 in total for "cits". We found that for 98 of 923 bundles a number in the column "pubs" is bigger than a number in the "cits". It means that for about 10% of cited papers there are citing papers without mentioning them in their in-text citations.

These confirmed the obvious fact: there is no sense to use the traditional citation index, if we know a number of mentions for citations. As wrote Pride and Knoth [17]: "Our results therefore confirm that the number of times a citation appears is a strong indicator of that citation's influence."

Co-citations. Co-citations that count how many times references are mentioned together in the same citation context. This indicator shows how exceptional is the cited paper. Many co-citations means that the paper is cited among many others paper and for citing authors this paper is one of many similar research outputs. If a paper has few co-citations, it means the paper has an exceptional character for citing authors.

We counted the co-citations within bundles and presented the numbers in the column "with" of the same table. We found a big diversity of values: about 10% of cited papers have no co-citations at all, and another 10% have co-citations in 10+ times more than their number of mentions ("cits").

Spatial Distribution. Spatial distribution of the in-text citations within papers' full text. Usually, spatial distribution means a position of in-text citations within IMRaD (Introduction, Methods, Results and Discussion) structure of citing papers [2, 3]. Occurrence of some in-text citations in the sections Method and Results, and another - in the sections Introduction and Discussion can mean that the first group of cited papers (cited in the Method and Results sections) are more important, since they were men- tioned by citing authors in relation with their producing of new scientific knowledge.

Since the citation data built in Cirtec do not provide information about papers' IMRaD sections structure we counted the spatial distribution over 5 equal fragments of papers' full text excluding the tail started with the "References" section. To have a compatible picture of such distributions, we built for each bundle an indicator "x/x/x/x/x" shows a number of in-text citations for a cited paper in each of 5 fragments of citing papers' full text. Such indicators are presented in the column "distribution" of the same table. We found that there are three obvious groups of cited papers which vary by their spatial distribution: (a) a group with more or less uniform distribution over 5 fragments; (b) a group with bigger number of mentions in fragments at the start and at the end of papers; (c) a group with bigger numbers in the middle of papers.

4.3   Qualitative Characteristics

Using a sets of citation context text strings available for bundles in the "raw data" we can build following qualitative indicators[13].

Common Phrases. Common phrases within the citation contexts, e.g. selected by the n-gram method[14]. One can analyze what phrases of 2 or more words are the most often repeated in citation contexts. Such phrases can help with identification of lexical cliché or the cue terms [18] used by researchers for expressing why they cited papers, what polarity has the citation contexts and what functions have the in-text citations [7].

Based on available citation contexts grouped for the cited papers we made experiments with selecting n-gram of 2, 3 and up to 6 words for each bundle. Using the n-gram method we processed two versions of the citation contexts: original and normalized. As expected we got more n-grams for original citation contexts, but n-grams for the normalized version had bigger number of repetitions. In total, the most of n-grams are phrases with two words (2-gram), we got much less of 3-gram phrases, even less of 4-gamm, and so on. Only single bundles have 5- or 6-gram phrases.

In the column "common phrases" of the table at http://cirtec.ranepa.ru/analysis/ we presented the first selected 2-gram for normalized citation contexts with two numbers in brackets: a number of repetitions of this phrase in a current bundle and in total for all bundles. A hyperlink here opens a page with a full set of selected n-grams for the current bundle.

Topic Models. Topic models (significant words or abstract topics) extracted from the citation contexts, e.g. by the Topic Model approach[15] and using the LDA method[16]. This qualitative indicator shows a set of the most relevant topics (contained each, e.g., 5 words) which reflect a thematic structure of citation contexts. Such information helps to understand what relations have the cited papers with thematical topics of citing papers. It also allows to compare a differences and commons between thematical topics of citing papers and cited papers [11].

---

[13] Aleksandr Tuzovsky and Amir Bakarov made needed software and calculations.

[14] https://en.wikipedia.org/wiki/N-gram.

[15] https://en.wikipedia.org/wiki/Topic_model.

[16] https://en.wikipedia.org/wiki/Latent_Dirichlet_allocation.

In the "topic models" column we presented the first extracted topic with a hyperlink. A page opened by the hyperlink contains all built topics with their relations to specific citation contexts measured by a probability coefficient.

Lexical Difference/Similarity. Lexical difference/similarity between citation contexts. One of possible approaches for such analysis is the word embedding[17] implemented as Word2Vec method[18]. The method is able to capture multiple different degrees of similarity between words [12].

He and Chen [9] used this method "to provide temporal representations of in-text citation of publications by word embedding models trained from citation context text, which can be used to characterize and analyze the changing and complex roles of the publications". Another scholars used Word2Vec to create the cite2vec method: "Our approach projects words that are representative of documents into a 2D space, and subsequently projects documents into this same space such that a document's proximity to a word indicates its manner of usage, while also preserving similarity of document-to-document usage" [1]. They wrote: "cite2vec visualizes documents and words in a way that adheres to such citation contexts, so that the user can explore and discover documents in a usage-oriented manner" [1].

For the citation content analysis this method allows to calculate a semantic distance not only between single words from citation contexts, but also between pair or groups of selected citation contexts. As a result we can measure similarity of citation contexts related with the same cited paper or with the cited author, etc. It gives us information about diversity of contexts where a paper or an author is cited.

For our experiments with this method we enriched the citation contexts dataset by adding ID of the citing paper into each citation context text string. We split up the ID on four part. It allows us to analyze similarities for following cases: (a) the selected citing paper; (b) all citing papers of the selected journal; (c) all citing papers of the selected publisher. This dataset is available at http://cirtec.ranepa.ru/Word2Vec/ (see files with the "txt" extension).

We made experiments with clustering of citation contexts belonged to the same bundles by their lexical similarity. In the column "dif" of the table at http://cirtec. ranepa.ru/analysis/ there are numbers of average "distance" between the bundle's citation contexts and the center of the bundle's cluster (centroid)[19]. The smaller the numbers, the more lexical similarity among the citation contexts exists within this bundle, and vice versa.

Hyperlinks in the column "dif" open a page with details about clusters built for each bundle. At the page there is a table with a list of built clusters for a bundle (a cluster with the number "−1" contains all non-clustered citation context) and a list of citation contexts belonged to each cluster.

---

[17] https://en.wikipedia.org/wiki/Word_embedding.

[18] https://en.wikipedia.org/wiki/Word2vec.

[19] https://en.wikipedia.org/wiki/Cluster_analysis#Centroid-based_clustering.

5 Conclusion

The Cirtec project provides for public re-use the open citation data and illustrates usefulness of these data for citation content analysis by experiments with building some numeric and qualitative indicators. These indicators can be easily linked with initial papers, authors of papers, etc. because they have inherited IDs. These results, if they are integrated with DL, can enrich traditional for DL representation of papers, authors and organizations.

Using the method illustrated by the Fig. 1 one can integrate these indicators into the Socionet digital library and visualize them as computer-generated annotations for appropriate fragments of papers in PDF. As a result, readers of papers in DL will get more useful information related to the citation content of the papers.

Acknowledgements. A part of this research – the approach of using citation contexts for building statistics with focus on the supercomputer simulation of interactions among the agents and research community environment, is funded by RSF grant (project No. 19-18-00240).

References

1. Berger, M., McDonough, K., Seversky, L.M.: cite2vec: citation-driven document exploration via word embeddings. IEEE Trans. Vis. Comput. Graph. **23**(1), 691–700 (2017). https://doi.org/10.1109/TVCG.2016.2598667
2. Bertin, M., Atanassova, I.: A study of lexical distribution in citation contexts through the IMRaD standard. In: Proceedings of the First Workshop on Bibliometric-Enhanced Information Retrieval Co-located with 36th European Conference on Information Retrieval (ECIR 2014), 13 April 2014, vol. 1143, pp. 5–12 (2014)
3. Bertin, M., Atanassova, I.: Factorial correspondence analysis applied to citation contexts. In: BIR@ ECIR, pp. 22–29 (2015)
4. Bertin, M., Atanassova, I., Gingras, Y., Larivière, V.: The invariant distribution of references in scientific articles. J. Assoc. Inf. Sci. Technol. **67**(1), 164–177 (2016). https://doi.org/10.1002/asi.23367
5. Bertin, M., Atanassova, I.: InTeReC: in-text reference corpus for applying natural language processing to bibliometrics. In: Proceedings of the Seventh Workshop on Bibliometric-enhanced Information Retrieval (BIR), Grenoble, France, pp. 54–62. CEURWS.org (2018)
6. Bilder, G., Lin, J., Neylon, C.: Principles for Open Scholarly Infrastructures. Science in the Open (2015). https://doi.org/10.6084/m9.figshare.1314859
7. Boyack, K.W., van Eck, N.J., Colavizza, G., Waltman, L.: Characterizing in-text citations in scientific articles: a large-scale analysis. J. Inform. **12**(1), 59–73 (2018). https://doi.org/10.1016/j.joi.2017.11.005
8. Ding, Y., Zhang, G., Chambers, T., Song, M., Wang, X., Zhai, C.: Content-based citation analysis: the next generation of citation analysis. J. Assoc. Inf. Sci. Technol. **65**(9), 1820–1833 (2014). https://doi.org/10.1002/asi.23256
9. He, J., Chen, C.: Understanding the changing roles of scientific publications via citation embeddings. arXiv preprint arXiv:1711.05822 (2017)
10. Hernández-Alvarez, M., Gómez, J.M.: Survey about citation context analysis: tasks, techniques, and resources. Nat. Lang. Eng. **22**(3), 327–349 (2016). https://doi.org/10.1017/S1351324915000388

11. Jebari, C., Cobo, M.J., Herrera-Viedma, E.: A new approach for implicit citation extraction. In: Yin, H., Camacho, D., Novais, P., Tallón-Ballesteros, Antonio J. (eds.) IDEAL 2018. LNCS, vol. 11315, pp. 121–129. Springer, Cham (2018). https://doi.org/10.1007/978-3-030-03496-2_14

12. Mikolov, T., Chen, K., Corrado, G., Dean, J.: Efficient estimation of word representations in vector space. arXiv preprint arXiv:1301.3781 (2013)

13. Parinov, S.: Towards a semantic segment of a research e-infrastructure: necessary information objects, tools and services. Int. J. Metadata Semant. Ontol. 8(4), 322–331 (2013). https://doi.org/10.1504/ijmso.2013.058415

14. Parinov, S.: Semantic attributes for citation relationships: creation and visualization. In: Garoufallou, E., Virkus, S., Siatri, R., Koutsomiha, D. (eds.) MTSR 2017. CCIS, vol. 755, pp. 286–299. Springer, Cham (2017). https://doi.org/10.1007/978-3-319-70863-8_28

15. Parinov, S.: Open citation data and a digital library. In: The Selected Papers of the XX International Conference on Data Analytics and Management in Data Intensive Domains (DAMDID/RCDL 2018), Moscow, Russia, 9–12 October 2018, vol. 2277, pp. 216-221. CEUR (2018)

16. Parinov, S., Lyapunov, V., Puzyrev, R., Kogalovsky, M.: Semantically enrichable research information system SocioNet. In: Klinov, P., Mouromtsev, D. (eds.) KESW 2015. CCIS, vol. 518, pp. 147–157. Springer, Cham (2015). https://doi.org/10.1007/978-3-319-24543-0_11

17. Pride, D., Knoth, P.: Incidental or influential?–a decade of using text-mining for citation function classification. In: 16th International Society of Scientometrics and Informetrics Conference, Wuhan, 16–20 October 2017 (2017)

18. Qayyum, F., Afzal, M.T.: Identification of important citations by exploiting research articles' metadata and cue-terms from content. Scientometrics 118, 21–43 (2019). https://doi.org/10.1007/s11192-018-2961-x

19. Waltman, L.: A review of the literature on citation impact indicators. J. Inform. 10(2), 365–391 (2016). https://doi.org/10.1016/j.joi.2016.02.007

Author Index

Printed in the United States
By Bookmasters

Printed in the United States
By Bookmasters